Teaching
Elementary
Statistics
with JMP®

Chris Olsen

The correct bibliographic citation for this manual is as follows: Olsen, Chris. 2011. *Teaching Elementary Statistics with JMP®*. Cary, NC: SAS Institute Inc.

Teaching Elementary Statistics with JMP®

Copyright © 2011, SAS Institute Inc., Cary, NC, USA

ISBN 978-1-60764-769-0 (electronic book)
ISBN 978-1-60764-668-6

All rights reserved. Produced in the United States of America.

SAS Institute Inc., SAS Campus Drive, Cary, North Carolina 27513-2414

1st printing, November 2011

SAS® Publishing provides a complete selection of books and electronic products to help customers use SAS software to its fullest potential. For more information about our e-books, e-learning products, CDs, and hard-copy books, visit the SAS Publishing Web site at **support.sas.com/publishing** or call 1-800-727-3228.

SAS® and all other SAS Institute Inc. product or service names are registered trademarks or trademarks of SAS Institute Inc. in the USA and other countries. ® indicates USA registration.

Other brand and product names are registered trademarks or trademarks of their respective companies.

Contents

About This Book

Purpose

The purpose of *Teaching Elementary Statistics with JMP* is to provide an introduction to JMP for teachers of elementary statistics. Just as statistics is data in a context, this book presents JMP in a context: teaching statistics. To accomplish this goal, I have chosen what I hope are examples of interesting data, and have interspersed JMP techniques and statistical analyses with thoughts from the statistics education literature.

Is This Book for You?

This book will be useful to all who are teaching statistics, but is specifically targeted at high school teachers of Advanced Placement Statistics and teachers of statistics in community colleges. Not infrequently, AP Statistics teachers and statistics teachers in community colleges are trained primarily as mathematicians, and are just beginning their journey into what is for them a new and exciting discipline. If you are beginning this journey, and even if you have progressed a bit, this book will help you with the tools, provide data sets for you and your students, and give you a sense of what teaching statistics and using great software is all about.

Scope of This Book

This book covers statistical techniques that would appear in a typical one-semester non-calculus based elementary statistics course. Specifically, the AP Statistics Course Description functioned as a gatekeeper for topics. Univariate and bivariate techniques are covered, as well as inference for means, proportions, slopes of regression lines, and categorical data.

JMP Typographic Conventions

As you proceed through this book, your attention will be directed back and forth from text to screen. The following typographical conventions will help you find your place in the text as you simultaneously attend to book, keyboard, mouse, and screen.

Style name	Description	Style specifics
body	This font is used for most text.	Times New Roman, 10.5 point
italic	The *italic* font is used to identify new terms.	Times New Roman, italic, 10.5 point
bold	The **bold** font highlights items of special importance.	Times New Roman, bold, 10.5 point
JMPChoice	The **JMPchoice** style is used to indicate a selection from a JMP menu. Such sequences display as phrases separated by a greater than sign, for example, **File → Save As**. This style is also used to identify variable names and data table names.	Arial, bold, 9.5 point
If you have set your JMP Preferences to something other than the default settings, some items might display differently on your screen.		

Software Used to Develop the Book's Content

This book was written using JMP 8 with Windows 7. Users of Macs or those with Linux operating systems should see only minor deviations from the look and feel of the presentation in this book.

Data and Programs Used in This Book

Data and programs used in this book can be found on the author's page at
http://support.sas.com/publishing/authors/olsen.html.

Author Pages

Each SAS Press author has an author page, which includes several features that relate to
the author including a biography, book descriptions for coming soon titles and other titles
by the author, contact information, links to sample chapters and example code and data,
events and extras, and more.

You can access this author's page at
http://support.sas.com/publishing/authors/olsen.html.

Comments or Questions?

If you have comments or questions about this book, you may contact the author through
SAS as follows:

Mail:

SAS Institute Inc.
SAS Press
Attn: Chris Olsen
SAS Campus Drive
Cary, NC 27513

E-mail: saspress@sas.com

Fax: (919) 677-4444

Please include the title of the book in your correspondence.

For a complete list of books available through SAS Press, visit
http://support.sas.com/publishing.

SAS Publishing News

Receive up-to-date information about all new SAS publications via e-mail by subscribing
to the SAS Publishing News monthly eNewsletter. Visit http://support.sas.com/subscribe.

x

Acknowledgments

It is patently obvious that a work such as this could not be imagined, let alone written, without the prior work of researchers, statisticians, and programmers. Nor would writing be pleasant were it not for the understanding of colleagues and friends, and the support of many who know far more than I about JMP, statistics, and writing.

The amateur computer programmer in me must acknowledge with awe the legion of individuals, unknown to me, who designed and programmed JMP. It is they who molded and shaped JMP into the powerful data analysis tool it is. Over the years, Lee Creighton and Mark Bailey have generously shared their knowledge of JMP. Curt Hinrichs, surely one of the best ambassadors of JMP one could imagine, has been incredibly supportive of this project (and incidentally knows the best seafood places in the Bay area).

With any large writing project, herding the prose of authors who are quite sure that a first draft is perfect, and that any draft after that is even more so, falls to editors. George McDaniel has been marvelously gentle with suggestions for improvement, and incredibly patient while I failed at retirement twice during the writing of this book, thus pushing back the initially anticipated completion date. Stacey Hamilton furnished much appreciated added value by superimposing some semblance of proper English form and style on my paltry scribbles.

The heart and soul of statistics is data. My aversion to made-up data makes frequent and lengthy trips to libraries a necessity. The fruits of such trips are the data sets that appear in this book. I must gratefully acknowledge the kind permission of authors, researchers, and publishers to reproduce these data sets. I trust that my presentations and interpretations of their data are faithful to the truths their labors have uncovered. If there be errors in presentation or interpretation, they are—of course—my own.

The vast lion's share of authorial abuse surely must fall on those closest, in this case my long-suffering wife. Over many years with an average of 4.3 "Dinner's ready! <Grrr>" calls per night, she has kept her equanimity, and not even once have I felt the sting of any of the sharp tools she uses for stitching and embroidery, though surely I have been very deserving. I hereby extend my condolences to all the men in the world: There can be only one Best Wife, and I have her.

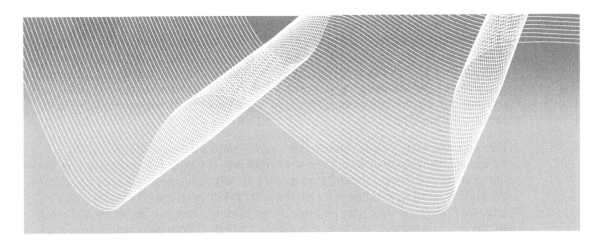

Chapter 1

An Introduction to JMP

An Introduction to the Introduction

My goal in this chapter is to quickly get you up and running with JMP, by paying special attention to those features you will most probably wish to use and by drawing your attention to some of the important features of the JMP interface. As we work through the introduction, I encourage you to experiment with the features of JMP at your leisure. Interspersed with the discussion about JMP, I present some results from the research on teaching and learning statistics and relate those findings to the use of JMP in the classroom. My intent is to get you past the beginner stage and into the productive stage with JMP; as it happens, the ease of use of JMP will make this an easy task. To advance to the "guru" level, you should every now and then pick up the materials that come with JMP and leaf through them. They are well written, and you will undoubtedly discover new gems of JMP capability as you read. I know how little time you have in your teaching day, but perhaps an opportunity for casual JMP study will present itself at the next faculty meeting.

I am aware of many people's aversion to reading software manuals. This is quite justifiable; manuals often seem to be written for people who (a) already know how to use the software or (b) wrote the software themselves. My fellow teachers and I are particularly adept at avoiding software manuals, a skill developed over many years of large class sizes and little time for professional reading. Only as a last resort do I reach for the software manual, and even then—I must admit—I frantically search for a quick fix. Manuals must communicate with users of all stripes, users with different levels of computer experience, a difficult task at best. Fortunately, my task here is much easier! This book is aimed at teachers of elementary statistics classes: a high-quality group with preexisting familiarity with computers.

The modern graphical user interface imposes a consistent look and feel across different software packages, and as a teacher with significant computer experience, I am sure you don't need nor do you want a lot of basic hand-holding about your computer's interface. That being said, there are features unique to statistical programs, and moreover features unique to JMP, that will extend your prior software experience. I will write in more detail when discussing these features.

As you know, statistical analysis is performed within the context of a problem, and the specific procedures used are defined by the nature of the statistical question and the available data. I realize that you may not have a ready stack of statistical problems complete with data sets, so I have included a few as examples. I include the problems and the accompanying data analyses to illustrate some of the capabilities of JMP, and especially its user interface. In addition, I hope that you will find the analyses interesting and usable in your classrooms.

At each step in the analyses to follow, JMP's interface gives you a sense that you are in control of the process. You, not the software, are performing the analysis. If you have struggled with other statistical software in the past, you will better appreciate this aspect of JMP. This sense of conversation with data, a conversation facilitated by the software, is a design feature, not an accident, of JMP. A typical data analysis using JMP begins with an initial presentation of graphs and statistics. Notice I said that the analysis *begins* with an initial presentation. We seldom know in advance exactly how we will analyze data until we get an initial look at it. JMP encourages that exploration and offers help and guidance at each step. You will see how all this happens as we proceed.

Some Stylistic Conventions

My natural writing and teaching style is to engage others in a conversation, rather than lecture. I have a general aversion to using the "I" word, and the authorial "we" sounds to me too much like the "royal" we. When I see "we" in a software manual I imagine Queen Victoria, in her best teacher voice, suggesting, "We are not amused." Generally, with first-person pronouns I will try to impart these meanings:

- "I" means I'm telling you what I am doing, what I'm thinking, or what I have experienced.
- "We" means that I'd like to guide you through something and am encouraging you to follow along. You are encouraged to experiment with JMP at any time, but if you do so when I am in the "we" mode your screens may differ from mine.

I will use basic terminology (click, drag, right-click, and so on) and am confident in your ability to master these directions. Frequently I will use the generic term "select," which usually means left-click. I am writing on a PC system, and the screen shots will be from a Windows 7 operating system. Presentation of JMP in a Mac or Linux environment is very similar to what you will see here.

Some of the look, feel, and behavior of JMP can be adjusted through choices made in the Preferences menu, which is found by selecting the **File** menu and clicking **Preferences**. The **Preferences** window is shown in figure 1.1. Rather than impose my own sense of what the preferences should be, I will continue to capitalize on the judgment of the designers of JMP. (In the many years I have used JMP, the only preference I have ever changed is the font.) As you become experienced with JMP you may find yourself repeatedly changing some aspect or other of JMP. When that happens, I encourage you to check out the options embedded in the Preference capabilities.

Figure 1.1 Preferences

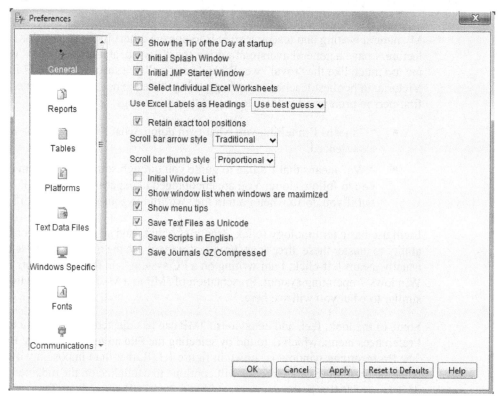

As we progress through the chapters I will be less detailed in my directions. As an example, in later chapters I will assume you know how to open a JMP file and will skip that step in the sequence of operations.

As you proceed through this book, your attention will be moving back and forth from text to screen. The following typographical conventions will help you find your place in the text as you simultaneously attend to book, keyboard, mouse, and screen.

- Reference to JMP entities will appear in the **Arial bold** font to cue you to look for those words in the presentation of JMP (for example, "Select the **File** menu, then the **Save** command" will be presented in syntactical English with the JMP words or phrases in Arial bold).

- When I ask you to select a sequence of commands, the sequence will appear as phrases separated by an arrow, such as **File → Save As**.

- I will present references to variable names, data table names, and some other items in **Arial bold**, but they may appear in illustrations in either a plain or boldface font. These items may also display differently on your screen if you have set your JMP Preferences different from the defaults.

- When there is a call for action on your part, please follow along with the example by following the instructions. "Hands-on" is an effective learning technique. Following the instructions will keep you from that quick-and-dirty strategy of imagining you know what would happen without actually doing it. (Been there, done that!)

One final ground rule: Please use the JMP Help features if you sense you might be bordering on being confused. JMP has an excellent Help system, and I will not feel slighted if you use Help to get a second opinion.

OK, let's begin. To start JMP, double-click the JMP application icon.

When JMP starts up you will see the JMP menu bar and the **JMP Starter** window (see figure 1.2). The **JMP Starter** window (the smaller one) is designed for the more experienced software user and presents functionality equivalent to the options presented above it. For our purposes we will close the **JMP Starter** window. (Closing the larger enveloping window will cause you to exit JMP.)

Figure 1.2 JMP Starter menu

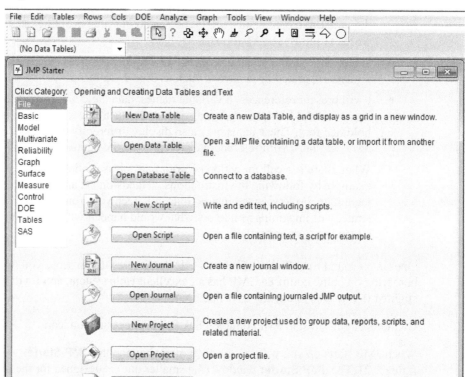

Consistent with other Windows applications, the **File** menu contains all the strategic commands, such as opening and saving data tables and exiting JMP. You may also see one of the "Tips of the Day," a set of nifty advisories about the capabilities of JMP. You can read them as they appear, run through the list of them, or even turn them off if you wish.

Creating a New JMP Table

This initial exercise will be the first and last time in this book that you will create a JMP table by entering data from scratch. After this data entry exercise you will open existing JMP data files. Statistical analysis of data is a context-driven activity, so for best results please read the context of the data as they are presented in sections and chapters to come.

It's a Cheetah-Eat-Gazelle World Out There!

For our first data analysis experience with JMP I will transport you to the grassland plains and savanna of the Serengeti in Tanzania, Africa. In the Serengeti National Park, cheetahs are predators of Thomson's gazelles. As a result, gazelles need to be attentive to the surrounding area, thus decreasing their vulnerability to attack, since one never knows what might be lurking nearby. However, gazelles feed on Serengeti grass and must therefore lower their heads to graze, thus increasing their vulnerability to attack. Cheetahs initially stalk their prey, generally moving to within 30 meters before they initiate the final chase. Upon reflection, it might seem reasonable that a fast cheetah could simply flush a herd of grazing gazelle and with a burst of speed pick off a slow gazelle for a fast lunch. Evolutionary development encourages energy efficient behavior, and one would think that a cheetah on stalk would be able to put off choosing its prey until the last moment and capitalize on a slow-moving element in a target-rich environment as the chase develops. However, once the cheetah initiates a chase it appears not to be dissuaded from an initially chosen target by the emergence of slower moving gazelles. This seems like a strategy that would burn excess energy. Is there an efficient predatory strategy that might account for this behavior?

Fitzgibbon (1989) has performed some research that may provide a possible answer. She reported on a possible relation between gazelle "vigilance" and the cheetah's choice of prey. In her study, she defined *gazelle vigilance* as the percentage of time during a stalk a gazelle had its head up scanning for predators. From a safe distance, she taped a series of stalks and kills by cheetahs and reviewed the footage to measure gazelle vigilance levels during the stalk. Fitzgibbon matched adult gazelles selected by the cheetah with the nearest actively feeding, same-gendered adult within five meters of them. She reasoned that if vigilance is a factor in prey selection, the vigilance of the selected victim should be less than the vigilance of the nearest neighbor who, after all, could have been chased just as easily. If vigilance is a factor, that could indicate the cheetah is "studying" the herd and identifying the less vigilant gazelles before settling on the specific prey.

Sixteen stalks' worth of data on the vigilance of the chosen gazelles and their nearest neighbors are presented in table 1.1. The data is of two types, quantitative (numeric) and qualitative (categorical). As you know, different statistical procedures are used with qualitative versus quantitative data. Mean- and median-based procedures are used on quantitative data, while frequencies and proportions are the usual fare for qualitative data. We will see as we proceed that JMP keeps these data types straight and will offer you appropriate statistical techniques during the analysis of data.

Table 1.1 Gazelle vigilance levels

Gazelle chased	Gazelle ignored	Gazelle gender
8.0	17.5	Male
31.4	63.0	Male
40.0	70.2	Male
15.7	38.8	Male
40.0	45.0	Male
35.3	39.1	Male
31.7	65.2	Male
68.5	72.5	Male
10.0	65.0	Female
78.7	84.2	Female
52.0	50.0	Female
23.9	20.5	Female
81.0	96.3	Male
49.7	90.1	Male
62.0	88.3	Male
72.6	89.9	Male
8.0	17.5	Male
31.4	63.0	Male
40.0	70.2	Male
15.7	38.8	Male
40.0	45.0	Male
35.3	39.1	Male
31.7	65.2	Male
68.5	72.5	Male
10.0	65.0	Female
78.7	84.2	Female
52.0	50.0	Female
23.9	20.5	Female
81.0	96.3	Male
49.7	90.1	Male
62.0	88.3	Male
72.6	89.9	Male
8.0	17.5	Male

1. If you have not already done so, close the **JMP Starter** window and select **File** →
   New → **Data Table**.

JMP initially displays data in a window in the form of a table with rows and columns
known as a **Data Grid**. In this window, you can perform a variety of table management
tasks: editing cells; creating, rearranging, or deleting rows and columns of data; and
sorting. Figure 1.3 identifies some of the commonly used parts of the window, including
the JMP data table. If you move the mouse over the table you will see the cursor change
shape now and then, alerting you to different functions that are potentially active when
the mouse is in a particular position. For example, if you position the mouse over one of
the contextual pop-up menus (the red triangles) the cursor changes to a finger-pointer,
indicating that you are over a clickable item. When you are at a boundary of a window
the cursor will change to a vertical, horizontal, or diagonal double-pointed arrow, alerting
you to the option of resizing the window.

Figure 1.3 JMP data table

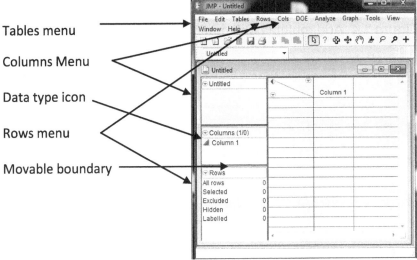

JMP data entry is similar to data entry in common spreadsheet programs. One significant
difference is that whole rows and whole columns, not individual cells, are conceptual
units of data entry. JMP tables consist of rows and columns, and JMP will interpret these
rows and columns of data, not individual isolated storage cells for numbers. An isolated
cell or group of cells can have user-defined meaning in a spreadsheet, but not in JMP.
Columns represent *variables* and rows represent *observations* or data values for the
variables in question.

We will need to create sixteen sets of three observations on these variables, and so we will need to create a data table consisting of sixteen rows and three columns. Variables can be added one at a time, or en masse. My own personal preference is to set up the table in advance before beginning data entry. Once the table is set up I enter the data a row at a time. It is not always possible to specify the number of rows in advance, since some data entry might take place before all the data are gathered. JMP allows rows to be added later (using **Rows → Add Rows**). Some may prefer to enter data in a manner similar to their method with calculators: enter observations one variable at a time. Still others just start entering, and the rows are updated on the fly. The "right" way to enter data for an individual is most likely whichever way results in fewer data entry errors. One method may take more or less time, use greater or fewer keystrokes, or perhaps even be more or less psychologically uplifting; the important value is getting the data entered correctly, not how it is entered. You should be guided by your sense of what is comfortable and natural for you. To initialize the data table for our cheetah data

2. Select **Cols → Add Multiple Columns**. JMP will display the **Add Multiple Columns** panel, as shown in figure 1.4.

Figure 1.4 Add multiple columns

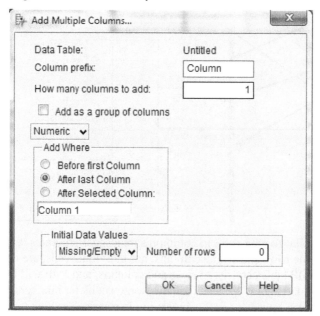

3. Respond to the **How many columns to add** question with "3".

4. Select **Rows → Add Rows**. JMP will display the **Add Rows** panel.

5. Respond to the **How many rows to add** question with "16".

Other options are available in the **Add Multiple Columns** panel, and you may wish to explore these. Our data consists of two quantitative variables and one qualitative variable, and we will specify the types of variables individually rather than **Add as a group of columns**.

6. Click **OK** and resize the windows so that you can see all the rows and columns.

Now that we have set up the data for table entry, let's actually enter the data. First, we will give the variables more informative names than **Column 1**. Figure 1.5 shows what the table looks like before we begin data entry. It is considered good form to enter descriptive variable names, and I will use the following: **Chosen, Ignored,** and **Gender**.

Figure 1.5 Initial columns and rows

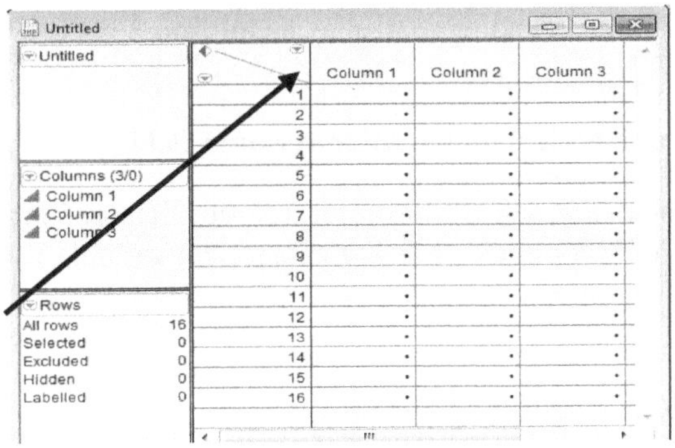

To edit the information in column 1:

1. Double-click on the column 1 heading to display a **Column 1** window, as shown in figure 1.6.

2. Enter the variable name (**Chosen**) in the **Column Name** field.

3. Click **OK**.

Figure 1.6 Set up a single column

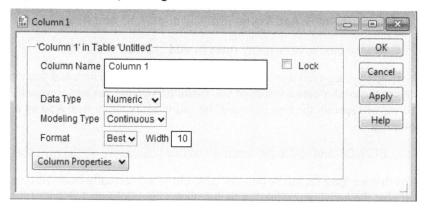

Now repeat this process and enter the information for column 2, using **Ignored** as the variable name. Column 3, **Gender**, will require a little more attention because it is a categorical variable, different from the default numeric data type in JMP.

1. Double-click on the column 3 heading to display a **Column 3** window, as shown in figure 1.7.

2. Enter the variable name (**Gender**) in the **Column Name** field.

3. Click the **Data Type** menu and select **Character**. (Notice that the **Modeling Type** changes to **Nominal**. JMP now knows this is a categorical variable.)

4. Click **OK**.

Figure 1.7 Set-up for a categorical variable

Now enter the first row of data. Be aware that on a PC with a numeric keypad, the Enter keys operate differently in JMP. The usual Enter key, when pressed, will send the cursor down to the next row. However, the numeric keypad Enter key, when pressed, will move the cursor across the columns before descending to the next row. If you wish to use the numeric keypad Enter key, you can change its behavior with the following keystroke sequence:

5. **File → Preferences → Tables → Numeric keypad Enter key moves down → OK**.

By whatever path and Enter key you choose, your first line of data should appear as shown in figure 1.8. Complete the data entry, check the numbers to make sure they are correct, and save the file with a name and in a location of your choice using the sequence:

6. **File → Save As**.

I will use "Gazelles" for the file name.

Figure 1.8 One line of data entered

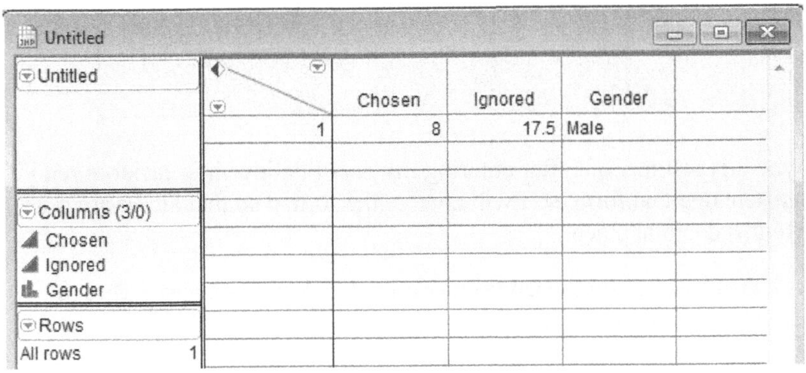

The JMP table should now look like figure 1.9 and reflect our progress so far. There are three columns (complete with variable names) and sixteen rows for data. The file name is Gazelles. JMP knows the **Chosen** and **Ignored** variables are quantitative, as evidenced by the small triangles showing in the **Columns** panel. Likewise, JMP knows that **Gender** is a character/nominal variable—as evidenced by the small bar-chart icon in the **Columns** panel.

Figure 1.9 Gazelles data

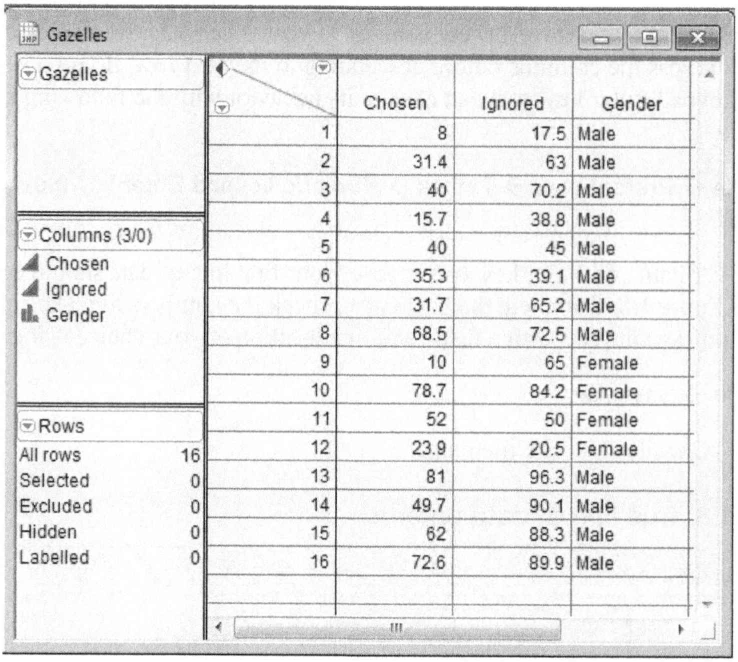

	Chosen	Ignored	Gender
1	8	17.5	Male
2	31.4	63	Male
3	40	70.2	Male
4	15.7	38.8	Male
5	40	45	Male
6	35.3	39.1	Male
7	31.7	65.2	Male
8	68.5	72.5	Male
9	10	65	Female
10	78.7	84.2	Female
11	52	50	Female
12	23.9	20.5	Female
13	81	96.3	Male
14	49.7	90.1	Male
15	62	88.3	Male
16	72.6	89.9	Male

Gazelles
- Gazelles
- Columns (3/0)
  - Chosen
  - Ignored
  - Gender
- Rows
  - All rows   16
  - Selected   0
  - Excluded   0
  - Hidden   0
  - Labelled   0

Maybe it is just me, but after entering my data I believe it is easier to read if it is in a consistent decimal form, so I will change the format so that all numbers are expressed to the tenths decimal place.

1. Double-click on the **Chosen** column heading to display a **Chosen** window, as shown in figure 1.10.

2. Select **Format → Fixed Dec**.

3. Select the **Dec** panel and enter "1".

4. Click **OK**.

Figure 1.10 Chosen window

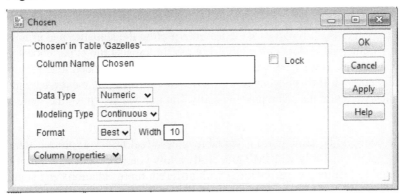

Repeat these steps for the **Ignored** column. Your completely entered data should now be displayed as shown in figure 1.11.

5. Select **File → Save** to make these changes permanent.

Figure 1.11 Completed data entry

	Chosen	Ignored	Gender
1	8.0	17.5	Male
2	31.4	63.0	Male
3	40.0	70.2	Male
4	15.7	38.8	Male
5	40.0	45.0	Male
6	35.3	39.1	Male
7	31.7	65.2	Male
8	68.5	72.5	Male
9	10.0	65.0	Female
10	78.7	84.2	Female
11	52.0	50.0	Female
12	23.9	20.5	Female
13	81.0	96.3	Male
14	49.7	90.1	Male
15	62.0	88.3	Male
16	72.6	89.9	Male

Gazelles

Columns (3/0)
◢ Chosen
◢ Ignored
📊 Gender

Rows
All rows 16
Selected 0
Excluded 0
Hidden 0
Labelled 0

Numbers and Pictures—Univariate Data

Now we get to the fun part of statistics: interpreting the statistics and graphs. We carefully entered the data in the table, but psychologists (and experience!) tell us that when we check as we type, we sometimes see what we intended to enter, not what we actually entered. If this is your first experience with JMP, take extra care to verify not only that we are in the same place visually, but that we also have the same numbers and pictures.

While being in the same place is valuable, please feel free to explore different JMP paths at any point—not only will JMP gracefully allow you to experiment, but it is also designed to facilitate experimenting with different representations and summaries of your data. The JMP folks know that data analysis is not a rigid lockstep process, but one of moving x steps in one direction and y steps in a different direction, depending on the unfolding story the data is telling. The data analyst ponders now and then, and sometimes takes paths less traveled by, depending on the data. If you go off on a tangent of exploration, it is easy to return to the main path we are traversing at the time.

I present a more detailed, more microstepped discussion in this chapter than in discussions that follow. You will see many important aspects of the look and feel of JMP in this example, and I will tend toward pointing them out in this chapter rather than in later chapters, when you will have more experience.

OK, let's analyze!

1. Select **Analyze → Distribution**.

2. In the **Select Columns** panel, click the **Chosen** variable.

You should see something similar to figure 1.12. (Your screen may vary a bit from mine depending on your preference for organizing windows.) You should not fret if you do not completely understand everything in this window. JMP is presenting many options to you because it does not yet know how simple or complex you intend your analysis to be.

Figure 1.12 Choosing the variable

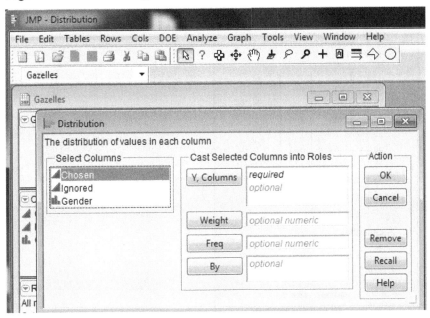

You will become very familiar with the **Select Columns** panel. It contains a list of the variables in the Gazelles file. The idea is that you will assign various "roles" to each variable. Many of the roles are for more advanced statistical analyses than you may be performing. For the most part the role selection in elementary statistical analysis will be choosing the variable(s) to analyze in a univariate setting, and choosing which variable(s) to be explanatory and response variable(s) in a bivariate setting.

3. Continuing on, key the sequence, **Chosen → Y, Columns → OK**.

Do you see those little icons I have circled in figure 1.13? They should appear as half-blue diamonds and downward-pointing red triangles. These are common and useful features in the various windows of JMP. The half-blue diamonds are "disclosure diamonds." Clicking on them opens or closes the associated item. You may choose to "disclose" or not the information associated with the icon. The red triangles, which JMP refers to as "hot-spot icons," present context-specific options for selection. These options are ones that are meaningful where you are in your analysis, that is, the "current context."

Figure 1.13 Distributions

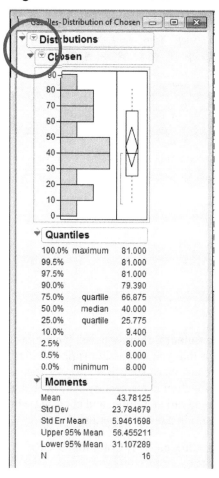

4. Click the disclosure icon beside **Distributions** to see its effect on the window.

See? The entire display was concealed. Now click on the same disclosure icon to expose it.

5. Click the disclosure icon beside **Quantiles**.

The parts of the display disappear and return as you wish—you can focus your attention on what is important to you at the time. This is also a nice feature when you want to transfer JMP output to your favorite word processor via copy and paste. JMP will transfer only the disclosed parts.

To repeat, the hot-spot icons offer different paths or a combination of paths, consistent with the context.

6. Click the **Distributions** hot-spot icon to see these further options:

    ▪ Uniform scaling
    ▪ Stack
    ▪ Script

If you select the **Stack** hot-spot icon the plots will display horizontally, as shown in figure 1.14. The **Chosen** hot-spot icon provides options specific to the **Chosen** gazelles variable.

Figure 1.14 Stack chosen

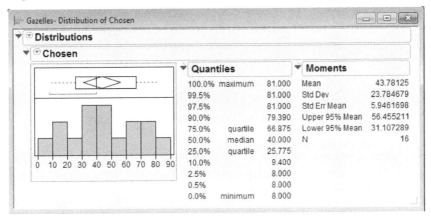

7. Select **Stack**.

8. Click the **Quantiles** and **Moments** disclosure icons to close them.

9. Click the **Chosen** hot-spot icon to display the options. You will see the **Display Options**, **Histogram Options**, **Normal Quantile Plot**, and so on.

10. From the list, select **Stem and Leaf** to add the stem and leaf plot to the histogram and box plot.

You should see a display similar to that in figure 1.15.

Figure 1.15 Some univariate options

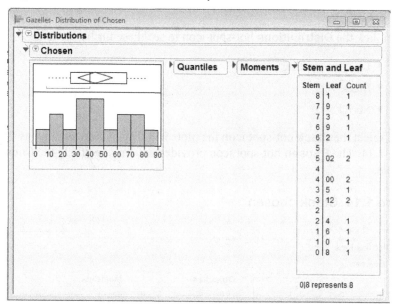

Before we proceed further, we will simplify our window somewhat.

11. Click the **Chosen** hot-spot icon.

12. Repeatedly select **Display Options** and deselect these options: **Quantiles**, **Moments**, **Stem and Leaf**.

13. Change the histogram and box plot to a comfortable size by clicking and dragging on the lower right corner of the box plot or the histogram.

Figure 1.16 displays my current window.

Figure 1.16 Histogram and box plot

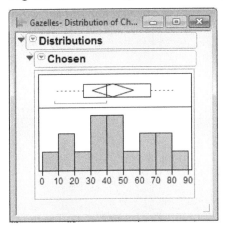

At this point we will experiment a bit and modify the graphs. You may have noticed that the box plot contains a diamond and a horizontal red bracket. Although these are informative, they are not box plot features we initially teach our students. The diamond presents the sample mean and a 95% confidence interval for the population mean. The horizontal red bracket indicates where the most concentrated 50% of the data are.

14. Right-click in the box plot area and deselect these options: **Mean Confid Diamond**; **Shortest Half Bracket**.

We will also change the appearance of the histogram. Move the mouse over the scale area of the histogram (as shown by the arrow in figure 1.17). Depending on where the mouse is, you will see a differently pointing right hand.

15. Left-click and drag with the mouse positioned at different positions in the scale.

You will quickly see the effects on the scale. You may have changed the bin size while experimenting with the scale, and some of the histogram bars may have a darker color, something like that shown in figure 1.17. Not to worry! We can undo that change.

Figure 1.17 After scale manipulation

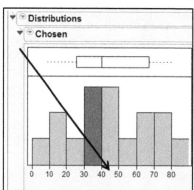

16. Double-click on the scale area (arrow in figure 1.17), and the **X Axis Specification** panel will appear as shown in figure 1.18.

Figure 1.18 X axis specification

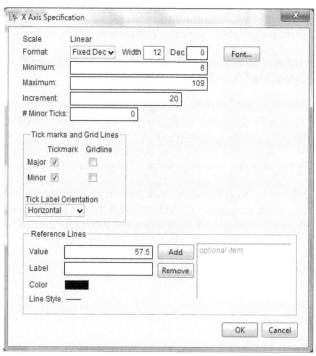

You can adjust various features of the scale in this window. (With bivariate data, there is also a **Y Axis Specification**.)

17. Change the **Minimum** to 0 and the **Maximum** to 109; change the **Increment** to 20, and click **OK**.

Figure 1.16 will reappear.

Thus far I have presented a short analysis of a distribution of a continuous variable. What about categorical variables? The procedures for a categorical variable unfold in a manner very similar to the quantitative variable. (This is another nice design feature of JMP: Learning is transferable.)

18. Select **Window** → **Close All Reports**.

19. Select **Analyze** → **Distribution** → **Gender** → **Y, Columns** → **OK**.

JMP presents a bar graph and a table in figure 1.19 with counts and proportions. I have used the **Distributions** hot-spot icon and selected **Stack** to change the orientation of the graph from vertical to horizontal.

Figure 1.19 Distribution of gender

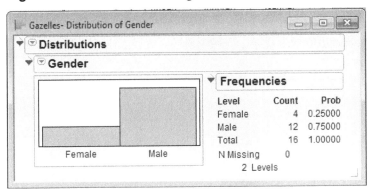

How did JMP know to generate a bar chart rather than a histogram? Recall that JMP recognized the **Modeling Type** of the variable as "nominal" (categorical) when you entered the data; JMP still remembers and once again presents the correct graph for the categorical data.

There is also an alternate, and probably better, procedure to display a bar chart.

1. Select **Window** → **Close All Reports**.

2. Select **Graph** → **Chart**.

3. Select **Gender** → **Categories, X, Levels** → **OK**.

Notice the **Chart** hot-spot icon. Let us see what sort of options we have there. We can get a **Horizontal [bar] Chart** and a **Pie Chart**, among others.

4. On the **Chart** hot-spot icon, select **Horizontal Chart.**

5. Then click again, this time on **Pie Chart**.

Before moving on to bivariate data, you may wish to experiment with options presented in **Y Options**, **Level Options**, and **Label Options**.

Here I make a brief but important teaching point about this presentation of the distribution of gender. I prefer what is shown in figure 1.20 for a bar chart. Experienced statisticians may be comfortable with the plot in figure 1.19, but it is something of a problem for those who are still learning statistics. The confusion between bar charts and histograms is mentioned often in research articles about students' interpretation of graphs. Frequently students will attempt to infer something about shape from a bar graph and even try to calculate means and standard deviations if the data are coded as integers. (See, for example, Bright and Friel [1998].)

Figure 1.20 A better bar chart

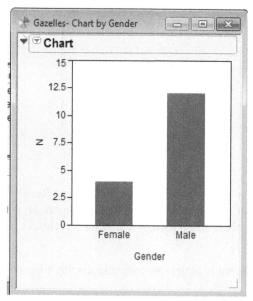

delMas, Garfield, and Ooms (2005) reported that even college students confuse histograms of numeric data with bar charts of categorical data. Students in this study felt they could use bar graphs of categorical data to estimate center, shape, and spread of a distribution. The graph in figure 1.19 is sort of a hybrid; it is a graph of categorical data but has the appearance of a histogram. Good information is presented nicely in this plot—but it can be something of a problem for students. A clear implication for teaching exists here. It is apparent that distinguishing categorical and numerical data is a topic that needs some attention. The explicit presentation of the data types in the data table in JMP supports the importance of making this distinction.

Numbers and Pictures—Bivariate Data

The premier bivariate method in elementary statistics is simple regression. Our two quantitative gazelle variables provide an opportunity to demonstrate a simple regression analysis in JMP. If you have not already done so:

1. Select **Window → Close All Reports**.

2. Select **Analyze → Fit Y by X**.

The **Fit Y by X** window will appear (figure 1.21). You will notice features of windows we have already seen as well as some new features.

Figure 1.21 Fit Y by X

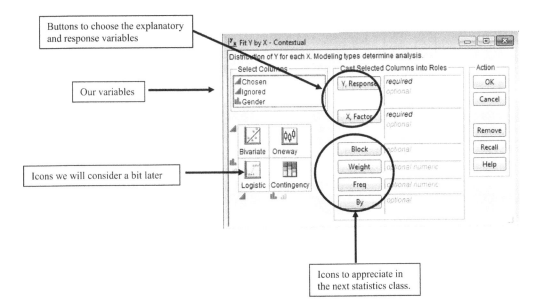

Now we must select our explanatory and response variables. We will choose the **Ignored** gazelle vigilance level as the explanatory variable, and the **Chosen** gazelle vigilance level as the response variable (see figure 1.22):

1. Select **Chosen → Y, Response**.

2. Select **Ignored → X, Factor**.

3. Select **OK**.

Figure 1.22 Chosen versus Ignored

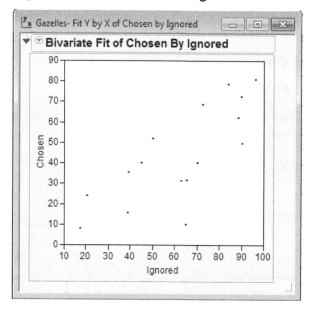

You may wonder why JMP uses the term *Factor* instead of *explanatory* or *independent* variable. This is because Factor is the term equivalent to "explanatory" that is usually used in the context of analyzing experiments; JMP is written to facilitate analyzing experiments as well as the analysis of data.

Notice the consistent presentation of graphs, especially the hot-spot icon. This hot-spot icon is where our bivariate analysis options will be presented—the options we need, and right when we need them.

Before proceeding, you might wish to revisit a JMP capability we have seen before. Check that you can modify each scale by using the right-hand icon, double-clicking on the axis areas. The points might be a bit small for some students, so we will enlarge them before adding the best-fit line to the scatterplot.

1. Right-click on the **Bivariate Fit** hot spot and select **Fit Line**.

2. Right-click on the graph area and select **Marker Size → 4, XL**.

3. Click on the graph area and select **Line Width Scale → 2.0**.

4. Click the **Analysis of Variance** and **Parameter Estimates** disclosure icons (the blue ones) to "undisclose" this information and thereby compact the **Fit Y by X** window.

The standard regression output is presented in figure 1.23. (You may need to enlarge your window to see it all.)

Figure 1.23 Best fit line

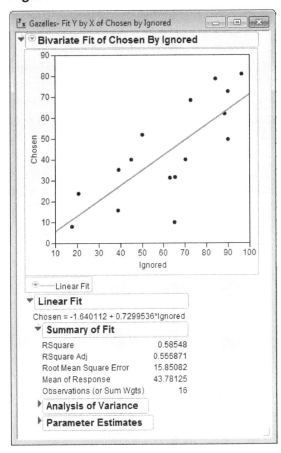

What Have We Learned?

In chapter 1 we learned the basic techniques of navigation with JMP and introduced the important contextual hot-spot and disclosure icons. We learned how to enter data and create files, as well as how to open and use existing files. Working through the gazelle data introduced the basic techniques of producing statistics and graphs in a univariate and bivariate analysis. It would not be too far afield to suggest that you already know 90% of how to generate univariate and regression reports commonly used in elementary statistics. In the next chapters, we will refine that knowledge, but at this point you are able to click your way anywhere and back with JMP.

References

Bright, G. W., & S. N. Friel. (1998). Graphical representations: Helping students interpret data. In S. P. Lajoie (Eds.), *Reflections on statistics: Agendas for learning, teaching, and assessment in K–12* (pp. 63–88). Mahwah, NJ: Lawrence Erlbaum Associates.

delMas, R. C., J. Garfield, & A. Ooms. (2005). *Using assessment items to study students' difficulty with reading and interpreting graphical representations of distributions.* Presented at the Fourth International Research Forum on Statistical Reasoning, Thinking and Literacy (SRTL-4), July 6, 2005, Auckland, New Zealand.

Fitzgibbon, C. D. (1989). A cost to individuals with reduced vigilance in groups of Thomson's gazelles hunted by cheetahs. *Animal Behavior* 37(3): 508–510.

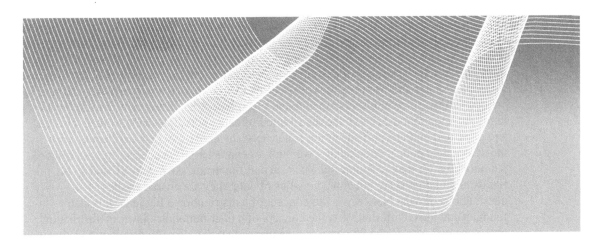

Chapter **2**

Distributions

Introduction

A detailed consideration of characteristics of univariate distributions occupies significant time and attention in elementary statistics. We use techniques of exploratory data analysis early in the formal study of elementary statistics to describe many distributions. The same techniques are used later to assess the credibility of the normality conditions that underpin inference with means, as well as the normality and homogeneity assumptions about errors in linear regression. In chapter 1 we saw how to enter data and display the default univariate display of information in JMP. In this chapter, I will delve more deeply into the power of JMP to analyze and compare univariate data distributions.

In elementary statistics, it is virtually impossible to overstate the importance of being able to examine distributions of univariate data easily and flexibly. A "statistical" response to data implies going beyond calculating measures of center. The conscientious data analyst must attend to the shape of a distribution of data, ask about the magnitude of the observed variability, and search for possible causes of that variability. We will explore data sets using a variety of techniques (stem and leaf plots, box plots, histograms, and so on) to highlight important aspects of the distribution of data, including shape, spread, clusters and gaps, and outliers. A distribution is a synergistic entity, not just a collection of individual data points. Graphically, we think of characteristics such as center and shape. Numerically, we may calculate statistics such as the mean, skew, spread, and so on. Statistics and graphs are not characteristics of data; they are characteristics of distributions of data. Not only is the concept of a distribution a big idea; it is also a fundamental idea. Unfortunately, the concept of a distribution is a level of abstraction above the data, and it should come as no surprise that the idea of a distribution is not necessarily a natural one for students.

In their review of the literature on the place of distributions in the statistics curriculum Garfield and Ben-Zvi (2008) point to "distribution" as one of the most important "big ideas" in statistics:

> Rather than introduce [distributions] early in a class and then leave it behind, today's more innovative curriculum and courses have students constantly revisit and discuss graphical representation of data, before any data analysis or inferential procedure. . . . [The] ideas of distributions having characteristics of shape, center, and spread can be revisited when students encounter theoretical distributions and sampling distributions late in the statistics course. (Chap. 8, p. 168)

Students learn the strengths and weaknesses of different representations of data by working with distributions in their various graphic forms. Box plots, for example, give a general idea of center and spread and are particularly useful for identifying outliers; stem and leaf plots and dot plots show more detail and may be used to spot clusters and gaps; histograms give a global sense of the shape of a distribution. Early in the learning process it is necessary for students to be able create graphs of distributions by hand. As these skills are honed, it is important for students to move beyond the construction of these graphs to actual reasoning about patterns in individual data sets, to comparisons between and among data sets, and finally to an understanding of what story the data have to tell. Students will acquire these data analysis skills in the same way one "gets to Carnegie Hall": practice, practice, practice. It is here that the utility of technology has a primary facilitative role. Fast production of graphs and statistics results in more time that can be spent on interpretation.

The statistical technology present in classrooms today consists of graphing calculators and computers. There are advantages to each. It is not my intent to rail against the use of graphing calculators, but rather to point out some advantages of JMP. Identifying the best mix of technologies is a decision properly left to individual classroom teachers and departments based on their unique circumstances, but it is beyond question that statistical software should be a major part of the mix. The advantages of JMP over the graphing calculator are impressive:

- Presentation quality of information is better with JMP. This is not simply a matter of more pixels on the computer screen: Consider also scales, axes, labels, units, and other information. Graphing calculators' failure to include this information with a graph raises the already high level of abstraction.

- JMP can provide varied representations of the same data on one screen.

- JMP can provide representations of different data sets in a single window for comparative purposes.

- Data entry is more reliable and simple with JMP; small calculator buttons and up and down arrows are more difficult to navigate than keyboards and mouse clicks.

- The keystroke overhead in changing graphic and/or numeric representations is relatively excessive with the graphing calculator; with JMP, the addition or deletion of a graph usually takes about two to three mouse clicks.

- Data transfer and student collaboration in the classroom and around the world is easier with JMP. JMP supports a wide range of data formats to facilitate data transfer as well as to download data from the Internet.

- JMP has a greater capability for working with large data sets.

- Finally, as we look to students' future college work and beyond, into their professional lives, the tools for serious work will be computers and statistical software. Teaching with JMP provides a better preparation for future college students and down the road supports their entry into the workforce.

Univariate Data

The data analysis problem described in this section comes from the work of anthropologist Margaret E. Beck (2002). She analyzed the tensile strength of ceramic objects from archeological sites in south central Arizona. Her unit of analysis was the "sherd," a fragment of pottery. Tensile strength is the maximum load that a material can support without fracture. Beck reported the load in units of force (kg) needed to shatter standard (equal area) sherd samples from her sites.

1. Select **File → Open**, and navigate to the folder in which you have stored this book's data files.

2. Select **TensileStrength → Open.**

JMP will respond by displaying the data table in figure 2.1.

Figure 2.1 TensileStrength data table

TensileStrength

TensileStrength

Notes C:\Documents and Se

	SherdSource	Weight (g)	MuThick(mm)	Load (kg)
1	Sacaton	14.7	5.41	15
2	Sacaton	14.6	5.75	30
3	Sacaton	20.2	7.37	51
4	Sacaton	15.7	5.75	20
5	Sacaton	15.6	5.70	17

The variable **SherdSource** indicates (no surprise!) the source of the sherd: Sacaton, Casa Grande, or Gila Plain, all sites in the Phoenix basin. The Sacaton and Gila Plain sherds date from AD 950–1100, and the Casa Grande sherds from AD 1100–1300. The variable **Weight (g)** is measured in grams, and **MuThick(mm)** is the mean thickness of the sherds measured in millimeters. The mean thickness is used because the surfaces of these sherds are slightly curved with variable thickness, typical of ceramic pots.

Now we will consider some of the univariate distribution options of JMP. Some of what follows will be a review of chapter 1.

1. Select **Analyze → Distribution.**

2. Select **Load (kg) → Y, Columns → OK.**

3. Click the **Load (kg)** hot spot; repeatedly select **Display Options** and deselect **Quantiles** and **Moments.**

4. Click the **Distributions** hot spot and choose **Stack.**

5. Repeatedly click on the **Load (kg)** hot spot and select **Normal Quantile Plot** and **Stem and Leaf.**

6. Right-click in the box plot area and deselect the **Mean Confid Diamond** and **Shortest Half Bracket** options, as we did in chapter 1.

You should see the data representations as shown in figure 2.2. You may wish to resize the graphics area by clicking and dragging in the lower right-hand corner of the histogram.

Figure 2.2 Common univariate displays

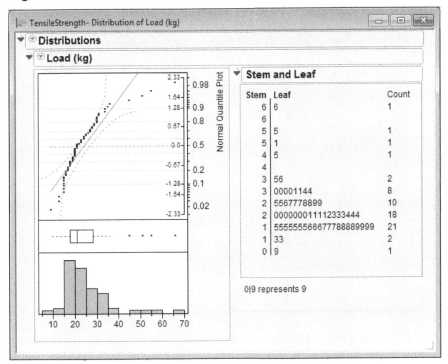

In their own different ways, each of the plots clearly indicates skewed data. The normal quantile plot may be less familiar to students than the box plot and histogram. If so, this may be an opportune time to introduce the excellent help capabilities built into JMP. (Remember, I encourage you to use the software's Help capability any time you feel the need for a bit more explanation.)

1. Select **Help → Index**.

2. Type in the keywords **Normal Quantile Plot** and click on **Display**.

3. Choose **Options for Continuous Variables** from the list and click **Display**.

A portion of the resulting Help information screen is shown in figure 2.3.

Figure 2.3 Normal Quantile Plot Help screen

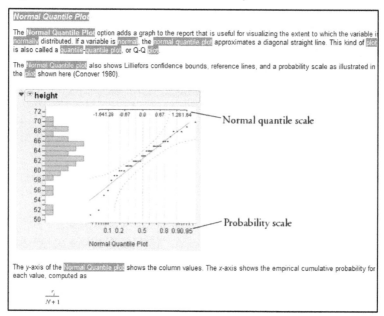

JMP has capabilities teachers and students can use to highlight the important aspects of graphs and focus student and audience attention on the most important concepts. As an example, consider the concepts of skew and outliers. Suppose that we wish to point out the different ways these plots indicate skew and outliers. We could annotate the graph to highlight or explain these important ideas.

1. Select **Tools → Annotate** and click-and-drag to create a text box.

Position the text box in a reasonable place on the graph by clicking and dragging; you may also have to make the text box wider and/or taller to display all of your text.

Key in an explanatory phrase such as "Indications of skew and outliers" in the text box. You can resize the box to suit your taste. Right-clicking inside the text box allows you to change some of its features, for example, the font.

2. Select **Tools** → **Simple Shape**; click and drag the ovals to the positions as shown in figure 2.4.

Figure 2.4 Annotated graphic information

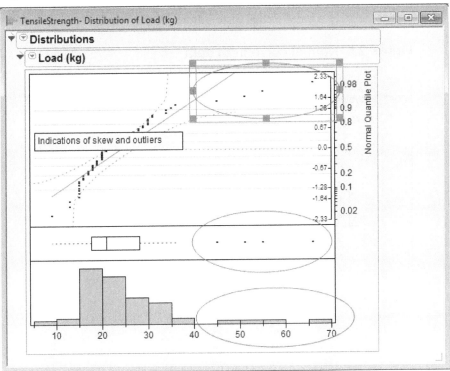

JMP allows you to easily tailor the presentation of your histogram to suit your personal taste. For example, by clicking on the **Load (kg)** hot spot, you have the option of removing or adding various graphical representations of the data. Try removing the box plot by deselecting **Outlier Box Plot**. (Repeat to remove all but the histogram.) Right-click the ovals that we created and select **Delete** to remove the oval shape. While we are here we can also implement a couple of perfecting amendments to our histogram presentation. If you wish, you may deselect the box plot and normal quantile plot in the display via **Load (kg)** → **Display Options**.

3. On the **Load (kg)** hot spot select **Histogram Options** → **Count Axis**. This adds a y-axis of observed counts to the histogram.

4. Again click the **Load (kg)** hot spot, this time selecting **Continuous fit** → **Normal**. This adds a normal curve to the graph.

You should now see a vertical axis similar to that in figure 2.5. The scale labeled "Count" presents the frequencies for each bar, and the normal curve associated with the sample mean and standard deviation is fitted to the data. The histogram's bars peek significantly above and below the curve, evidence of a lack of approximate normality in the data. This can be a further aid in contrasting the normal curve model with the usually "approximately normal" actual distributions we see with real data.

Figure 2.5 Normal curve fit to data

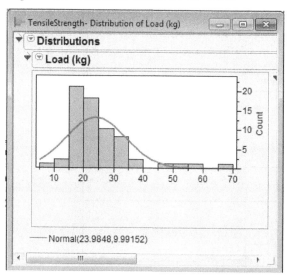

Comparing Distributions of Data

Once students are comfortable—or at least competent—with distributions, the next data analysis step involves comparing distributions; that is, comparing their characteristics to identify similarities and differences between and among their distributions. Comparing distributions of data sets is not just a task of analyzing one distribution multiplied by the number of distributions. Students who simply construct separate lists of characteristics of distributions when asked to compare them are missing something essential in their analysis. Students should go beyond analyzing the nature of the variability within each set of data, and also identify and comment on any differences in variability, shape, or other distinguishing features. The comparison of distributions at an early exploratory level in students' experience sets the conceptual stage for future comparisons using formal statistical inference. Beyond a statistics course, the statistical structure of most research studies reported in the press and increasingly available online is a comparison of distributions and their statistics. From a teaching standpoint, comparisons also lead to much more interesting class discussions.

Two prevalent graphical techniques useful in comparing distributions of data are histograms and box plots. Histograms allow a more detailed comparison of the shape of distributions; box plots allow a more condensed presentation of more than two batches of data. Of course, the relative advantages of histograms and box plots will vary with the nature of the data and the sample sizes.

To illustrate the power of JMP in comparing data we will once again consider the tensile strength data of Beck (2002).

1. Select **Window → Close All Reports**.

2. From the JMP data table, select **Analyze → Distribution**.

3. Select **Load (kg) → Y, Columns**.

4. Select **SherdSource → By → OK**.

This sequence tells JMP to display the **Load (kg)** for each **SherdSource** group separately.

5. Click the **Distributions** disclosure icon and choose **Stack**.

Choosing **Stack** will result in a display of the information for each sherd type on its own horizontal row.

6. Click the **Distributions** disclosure icon again, and this time choose **Uniform Scaling** for each **SherdSource** graph.

The **Uniform Scaling** choice will result in equal-length scale intervals across the three groups of data. That is, intervals of size 5 will appear with the same length on the screen irrespective of the **SherdSource**. (I have silently hidden the means diamonds and the most concentrated intervals.)

Your results should now be similar to those shown in figure 2.6, except you will see presentations of all three sherds, rather than just two.

Figure 2.6 Comparing distributions

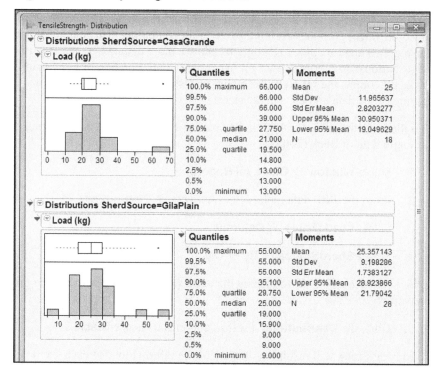

Some subtleties in the process of comparing histograms still might evade the attention of the unwary student comparing these groups of data. For example, the three scales range over different intervals. JMP has presented these graphs, reasonably enough, in a manner that maximizes the detail, given the area allocated on the screen. In the **Stack** representation, the capability of visual comparison is enhanced if the bin sizes and intervals are the same. Inspection of the data (by appealing to the stem and leaf plot or checking the quantiles) reveals that the smallest load in any of the sherds was 9 kg and the largest 66 kg. Considering this, it seems convenient to fix the endpoints of the intervals at 0 and 70 kg. I will arbitrarily suggest 10 kg for the increment, and throw in one minor tick per 10-kg interval. For each of the distributions, execute the following sequence:

1. Double-click the scale area to bring up the **X Axis Specification** panel.

2. Choose 0 for the **Minimum** field, 70 for the **Maximum** field, 10 for the **Increment** field, and 1 for the **# Minor Ticks** field. Leave the other settings at their default values.

The histograms should appear as shown in figure 2.7, except you will again see presentations of all three sherd sources. As usual, feel free to experiment with different values of axis settings to see their effects on the histograms.

Figure 2.7 Comparisons redux

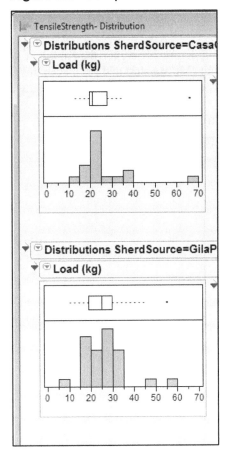

Pause for a moment to compare the presentations shown in figures 2.6 and 2.7. Creating uniform scales facilitates the quick visual comparison of distributions. The shapes of these distributions appear to be different; it now seems that the data from Sacaton and CasaGrande are skewed and GilaPlain's data is not. Greater detail can be displayed with the normal quantile plot.

For each of the sherd source sites:

1. Select **Load (kg)** → **Display Options**.

2. Select the **Normal Quantile Plot** and deselect **Quantiles, Moments**, and the **Outlier Box Plot**.

3. Select **Histogram Options** and deselect **Histogram**.

The display should look similar to that in figure 2.8. Again, the point here is that the route from original to alternative display using JMP was fast and easy, even this late in the analysis.

Figure 2.8 Normal quantile plots

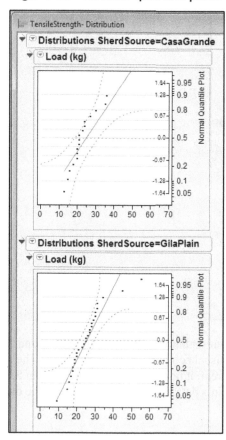

Since in elementary statistics the normal quantile plot is mostly about the shapes of distributions, we need not ruthlessly enforce uniform scaling. If you are unfamiliar with the normal quantile plot, it is instructive to examine these three plots simultaneously. From inspection of the histograms it appeared that the loads from the GilaPlain plot were closest to approximately normal; the two histograms for CasaGrande and Sacaton are skewed to the right. The box plots tell the same story, with the box plot for GilaPlain appearing to be more symmetric, showing whiskers of similar lengths. Not surprisingly, the normal quantile plots tell a consistent story. The plot for GilaPlain is closest to being straight by a larger margin, and the normal quantile plots for CasaGrande and Sacaton are similar to each other.

JMP also provides an alternative path to displaying multiple distributions.

1.   Select **Windows → Close All Reports**.

2.   Select **Analyze → Fit Y by X**.

3.   Choose **Load (kg) → Y, Response**, and **SherdSource → X, Factor → OK**.

The **Fit Y by X** choice might sound a little odd; perhaps some explanation is in order. We usually think of "fitting" in the context of regression. The dominant paradigm of statisticians today is the construction of models for Y, or response variables as functions of the X, or explanatory variables, and the generic term for this is "model fitting." The explanatory and response variables in the model can be either categorical or numeric. These models contain many of the techniques of elementary statistics under one regression umbrella, and JMP mirrors this by locating many techniques under the **Fit Y by X** moniker.

JMP keeps everything straight in the Fit Y by X world by checking the data types. It then offers options that are relevant for those data types, performs the appropriate procedures, and displays appropriate graphs automatically. JMP remembers that for our data the "response" variable is **Load (kg)** and the "explanatory" variable is **SherdSource**. The technical name for analyzing models with a quantitative response variable and categorical explanatory variable is one-way analysis of variance (**Oneway Analysis**). What we get out of all this is the graphic display in figures 2.9a and 2.9b.

Figure 2.9a Oneway ANOVA

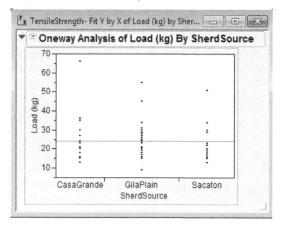

Figure 2.9b Oneway ANOVA with box plots

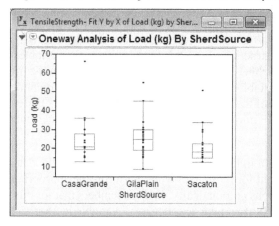

The **Grand Mean** is the horizontal line in figure 2.9a. It is positioned at the mean for all the **Load (kg)** data and functions as a line of reference much like the zero line in a residual plot in simple regression. The **X Axis Proportional** selection (which we have suppressed) allocates horizontal space in proportion to the sample sizes.

1. Click the **Oneway Analysis of Load (kg) By SherdSource** hot spot.

2. In the **Display Options**, deselect **Grand Mean** and **X Axis Proportional**.

3. In the **Display Options**, select **Box Plots**.

Your efforts will be rewarded by the transformation of figure 2.9a into figure 2.9b.

Beck was interested in whether there were technological changes over time in production that made the ceramic materials better able to withstand higher forces. The later CasaGrande site does appear to have higher load measures than the Sacaton site, suggesting an increase in strength. Beck asserts that this evidence is consistent with the fact that the CasaGrande vessels are less fragmented than the Sacaton vessels. Interestingly, the GilaPlain site, though dating from the earlier time period, appears to have the strongest sherds. Notice that the data points are presented with the box plots. This is optional, and the presentation of the data points can be deselected in JMP (**Oneway Analysis hot spot → Display Options**). Some discussion in the literature (Garfield and Ben-Zvi, 2008, chap. 8) suggests that students learning to analyze distributions should make a transition from seeing data as distinct points to seeing a distribution as a single entity. Graphic displays with both points and box plot may aid students in making this transition.

Finally, I would like to point out that the **Oneway Analysis** hot spot includes an option for showing histograms also.

1. Click the **Oneway Analysis of Load (kg) By SherdSource** hot spot.

2. Select **Display Options → Histograms**.

The resulting figure 2.10 presents a view at three different levels of detail: individual points, histograms with selectable bin sizes, and box plots.

Figure 2.10 Three views

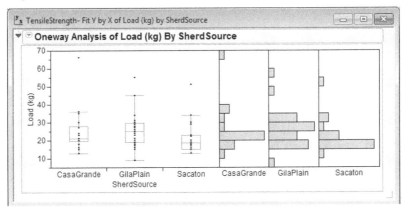

What Have We Learned?

In chapter 2, I discussed some of the advantages of JMP over graphing calculators. You learned how to annotate graphs and in general delved deeper into the capabilities for analyzing univariate data in JMP. We multiplied these capabilities when we considered how to compare groups of univariate data, a great leap forward for students conceptually. We also considered some teaching issues related to univariate plots. In chapter 3, I will illustrate some of these techniques with larger data sets.

References

Beck, M. E. (2002). The ball-on-three-ball test for tensile strength: refined methodology and results for three Hohokam ceramic types. *American Antiquity* 67(3): 558–69.

Fitzgibbon, C. D. (1989). A cost to individuals with reduced vigilance in groups of Thomson's gazelles hunted by cheetahs. *Animal Behavior* 37(3): 508–510.

Garfield, J., & D. Ben-Zvi. (2008). *Developing students' statistical reasoning: Connecting research and teaching practice*. New York: Springer.

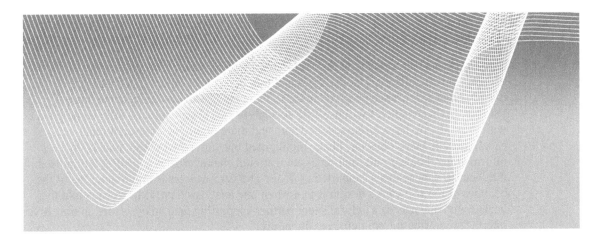

Chapter 3

The Analysis of Data—Univariate

Introduction

When we teach elementary statistical techniques we frequently use small data sets—for very good reasons. With small data sets, statistics such as the mean, interquartile range, and so on are quickly and easily calculated by hand, allowing students to gain an understanding of the processes behind the statistics, as encapsulated by the formulas. Unfortunately, small data sets are frequently dangerous data sets; not much insight can be gained by studying the distributions. For example, changing the bin size of a histogram of a small data set is often invites student confusion. Which of the differing histogram shapes is the "real" one? With scatterplots based on a small number of points, are we imagining curvature in a residual plot, or are we reaching into a random scatter of stars and creating a Big Dipper? Little insight into a population can be gained by examining small data sets.

Large data sets better facilitate the exploration of distributions and scatterplots and allow students to better discern stable shapes and trends. With advances in technology, downloading and working with large data sets in the classroom has become more feasible. In this chapter, I will use JMP to analyze two large data sets, offer some practice with the skills we have discussed, and here and there highlight some additional capabilities of JMP. Thus far, I have highlighted the various windows, panels of information, choice points, and other navigational characteristics of JMP. I will now shift the focus to data analysis supported by JMP. The verb *support* is important. One of the beauties of the software's design is that when one analyzes data, JMP recedes into the background and acts as a helpful silent partner supporting data analysis, not an obtrusive interface requiring intermittent user attention.

Espionage in the American Civil War

Data analysis begins with questions presented in a context. The geographic and historical context for our first leap into a large data set is the great state of Virginia during the early stages of the American Civil War, in 1861. The Confederate army system was modeled after the Federal system; units of soldiers known as companies were formed in counties and sent to their state capitals. There the companies were combined into larger groups called "regiments," almost exclusively comprised of ten companies. Many of these state regiments were sent to Richmond, the capital of the Confederacy, for service outside their state in the Confederate army as dictated by the fortunes of war.

At this time, General George McClellan commanded the Union forces in the eastern theater. An essential part of his planning depended on the number of Confederate soldiers his troops would face in any operations in Virginia. This parameter was, of course, unknown to the Federal forces. Random sampling not being an option, such estimation would normally be done by gathering information from observations in the field—that is, from spies. General McClellan's training at West Point did not include study in the art of espionage, so he asked Allan Pinkerton, the famous Chicago-based detective, to gather information about Confederate troops, and specifically to estimate their numbers and locations in Virginia.

In 1861 the editors of the northern press felt that McClellan should be marching to Richmond, engaging the Rebels, and putting an end to the insurrection. Conventional military wisdom was that "enough" men to successfully advance on entrenched forces was at least three times the number of troops available to the defenders. Pinkerton's estimate of the number of Confederate troops across the Potomac in neighboring Virginia was in excess of 100,000. General McClellan, in accordance with the aforementioned military thinking of the time, was not about to sally forth into battle until he achieved a more favorable ratio of attacking to defending troops than he currently had. This reluctance put him at odds with the press, the public, President Abraham Lincoln, and eventually military historians.

In point of historical fact, Pinkerton's numbers were wrong, and no one knew this better than the Confederate generals, whose memoirs began to appear in the 1880s. By 1885, historians were busily reviewing these personal accounts as well as available wartime documents and concluded that Pinkerton should have stuck to detective work. The only question of any interest was the extent to which McClellan was culpable for allowing himself to be misled by Pinkerton.

Our large-scale statistical analysis will focus on the question of whether Pinkerton's *method* of estimation, as described by him, *should* have worked. Pinkerton's first known report to McClellan, dated October 4, 1861, contains a clear estimation methodology in what we would now think of as a spreadsheet. He began by assuming that Confederate infantry regiments consisted of 700 men. After identifying the number of regiments then in Virginia he multiplied that number by 700. Then he subtracted 15 percent for sickness to arrive at his final estimate of troop strength. Pinkerton's determination of the number of regiments per state is not considered here. For a more complete analysis of Pinkerton's method, see Olsen (2005). It turns out that Pinkerton made a serious error of arithmetic in his report; Pinkerton (or his secretary) actually subtracted one-fifteenth, not 15 percent, of a regiment for sickness. Such an error would, of course, result in an inflated estimate. Our statistical question is not about the arithmetic; rather, our interest is in Pinkerton's method as presented in his report. Were his assumptions about the sizes of regiments and the level of sickness consistent with the facts on the ground at the time? We shall use JMP to help answer this question.

Data on company sizes, as well as numbers of individuals sick, were recorded in bimonthly Confederate company muster rolls. Recall that in the Confederate army the regiments were almost invariably made up of ten companies, so the leap from company sizes to regimental sizes is a shift of a decimal point. Data from the surviving Confederate company muster rolls for the months of June, August, October, and December 1861 are in the file Pinkerton.jmp.

1. Select **File** ➔ **Open** and navigate to wherever you have elected to save the JMP files.

2. Select **Pinkerton** ➔ **Open**.

JMP will respond by displaying the raw data table. The variable names appear in the **Columns** panel to the left of the data table, as shown in figure 3.1.

Figure 3.1 Muster roll variables

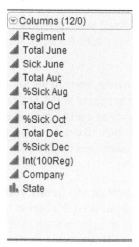

The first variable, **Regiment**, compactly encodes the state where the regiment was formed, the regimental number, and the company, A through J. As an example, 719.02 indicates company B (.02) of the nineteenth regiment in state number 7: Mississippi. From this encoding the company, regiment, and state have been reconstructed using some capabilities of JMP that we will see later. The variables of interest to us as we analyze the efficacy of Pinkerton's methods are the company sizes (**Total**) and the percentage of soldiers who were sick, injured, or otherwise not available for medical reasons (**% sick**). A quick glance at the data reveals that there are many cells with missing data. One reason for missing data is timing. If a regiment did not form until September it would not appear in these muster rolls until October. Another reason for missing data is that the collection of muster rolls is incomplete, some rolls having been lost in the fog of war.

Our analysis will focus on Pinkerton's basic assumptions: (1) there were 700 men per regiment, and (2) 15 percent of them were sick. The unit of analysis is the company, so we are assessing whether 70 soldiers per company and a 15 percent sick rate are consistent with the true conditions on the ground in 1861. Since the August muster rolls would be the most recent bimonthly report prior to Pinkerton's October 4th report to McClellan, we can focus our analysis on the August data.

1. Select **Analyze → Distribution → Total Aug → Y, Columns → OK.**

Arrange the layout of the graphs and numeric information to suit your preferences; my preferences are usually to display in a horizontal format (**Distributions → Stack**) as shown in figure 3.2.

Figure 3.2 Distribution for August 1861

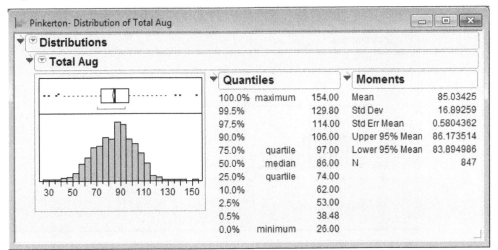

2.   Click the **Total Aug** hot spot and select **Continuous Fit → Normal.**

We see immediately in figure 3.3 that the August 1861 distribution of company sizes is approximately normal. It is interesting to note that, contrary to the accusation of estimates writ large, it appears Pinkerton's assumption of regimental sizes was actually too low. The mean in August was about 15 men per company higher than Pinkerton's assumption of 70 per company. The Confederate legislature had mandated that companies be no smaller than 64 men. With the benefit of hindsight, one might argue that since the companies were quickly formed and sent immediately for training after formation, adding a 10 percent "fudge factor" would give a reasonable estimate of company size. However, it is unknown where Pinkerton derived this number for company sizes, and there is nothing to indicate Pinkerton used that reasoning to arrive at such a fudge factor.

Figure 3.3 Normal fit

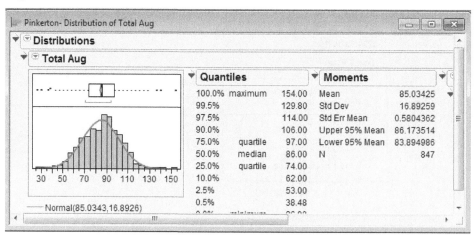

What about the proportion of men sick? Returning to the data table,

1. Select **Analyze → Distribution → % Sick Aug, Y → Columns → OK**.

The distribution of **%Sick** in figure 3.4 is positively skewed, with a mean 21.7 percent and a median of 19.5 percent. Pinkerton's assumption of 15 percent appears to be too low. His estimate of the number of "effectives" per regiment would be 85 percent of 700, or about 600 soldiers. The data for August suggests the mean number of effectives would be approximately 78.3 percent of 85, 66.6 per company, or about 660 effectives per regiment. Pinkerton's procedure, applied to August data, appears to be about 6 percent too high. That performance, while perhaps not great, is certainly not a searing indictment of incompetence either. How does Pinkerton's method in his October 4th report stack up for each of the four sets of the muster rolls? Once again, **Close All Reports**, and proceed to Step 1.

Figure 3.4 Distribution of %Sick, August 1861

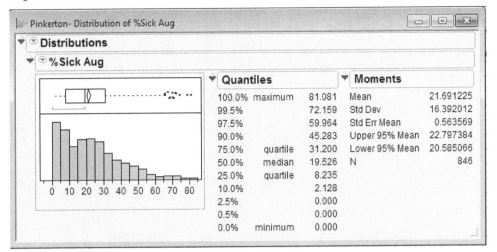

1. Select **Analyze → Distribution**.

2. While pressing the **Control** key, select all four of the **Total** columns (June, Aug., Oct., and Dec.).

3. Select **Y, Columns → OK**.

4. Click **Distributions → Uniform Scaling** and then again on **Distributions → Stack**.

5. For each histogram, click on the scale and make the following changes: **Minimum** = 0, **Maximum** = 160, **Increment** = 10, **# Minor Ticks** = 1.

The resulting distributions of the company totals shown in figure 3.5 represent a certain consistency of totals across these dates, suggesting that the August agreement between Pinkerton and regimental reality was not a simply a statistical fluke.

Figure 3.5 Company totals for four months, 1861

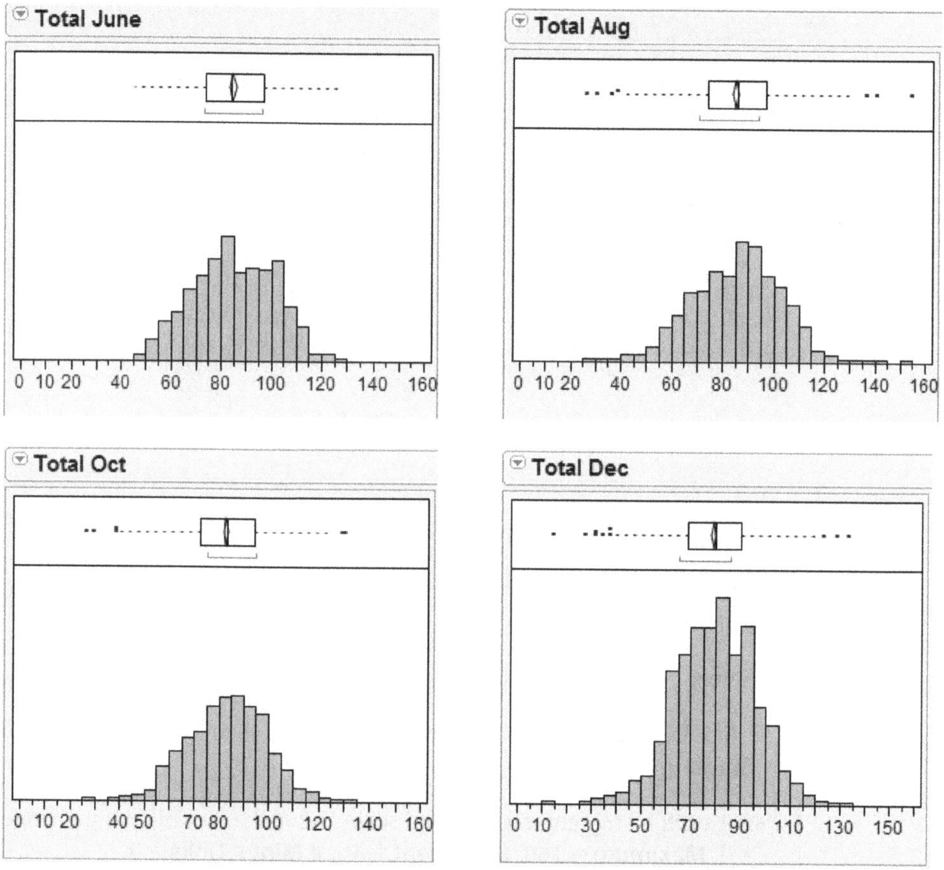

How does the **%Sick** compare across months? I encourage you to try this one on your own (don't forget the uniform scaling). You should see displays similar to that shown in figure 3.6.

Figure 3.6 Company %Sick for four months, 1861

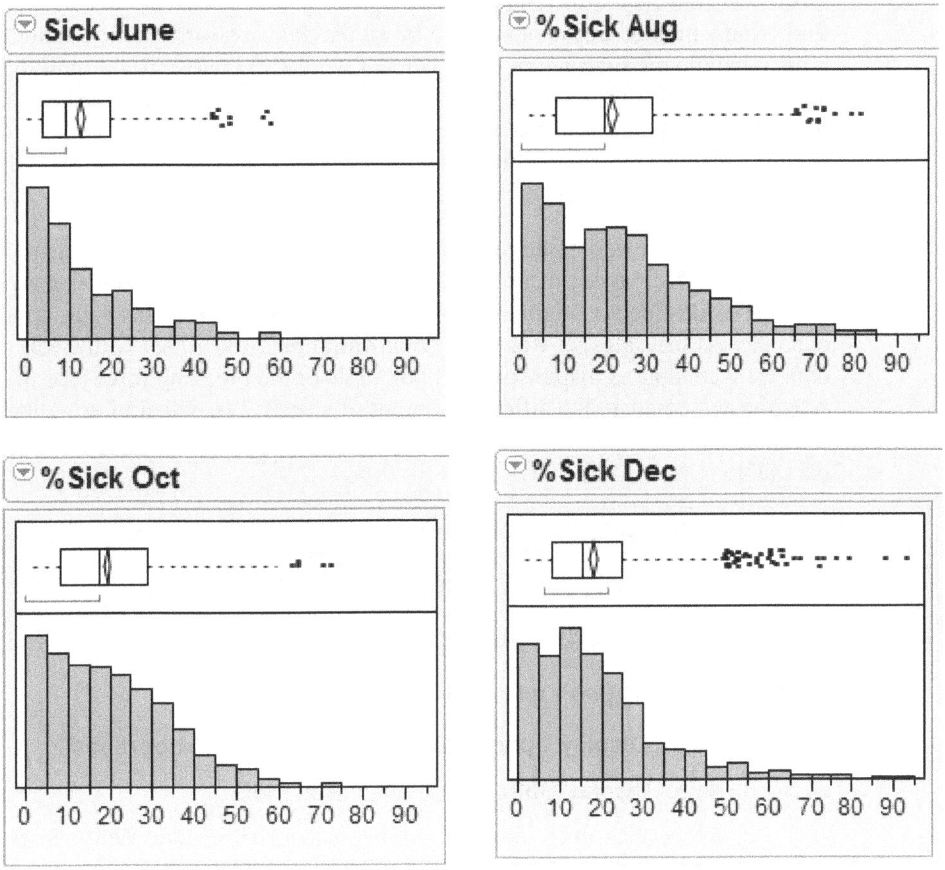

From these data it appears that while the shapes of the distributions are similar, the **%Sick June** is different from August, October, and December. The means for all companies in the sample are 12.8, 21.7, 19.5, and 18.3 percent, respectively. What might account for this pattern of **%Sick**? Notice that the number of companies for which we have muster rolls at the end of June was 47; then we have 84 at the end of August, 100 at the end of October, and 111 at the end of December. Perhaps the greater number of companies packed into the camps contributed to the spread of sickness. Perhaps it just takes a bit of time for sickness to spread. The first major battle of the war had been fought in July at Manassas Junction, Virginia, and after that there was a long lull in major action. Perhaps the 21.7 percent peak in August also reflects soldiers wounded in the July action.

The distributions of the company sizes and the **%Sick** over these six months (June to December) have unearthed a possibly surprising result: Contrary to historians' beliefs, Pinkerton's method does not seem to be all that bad. The arithmetic misunderstanding of how to handle the subtraction of 15 percent would, of course, result in higher estimates of numbers of "effective" troops. (Perhaps Pinkerton could have used a math teacher? There was a math teacher in the war at this time—Thomas "Stonewall" Jackson, lately of the Virginia Military Institute—but he was already serving in a different army for a different commander, General Robert E. Lee.)

Now that we have addressed our original questions about Pinkerton's methods and have the power of JMP at our disposal, we will do a little gratuitous snooping around in the data. One question we might ask of the data is, were the company sizes approximately equal across the states, or was there some variation? Recall that with Beck's (2002) sherd data we were able to display parallel box plots of the breaking forces for the sherds from Arizona and compare the different sources of sherds. We will similarly disaggregate the totals data and display the October company sizes by state. Once again, return to the JMP data table.

1. Select **Analyze → Fit Y by X → Total Oct → Y, Response**.

2. Choose **State** for the **X, Factor** and click **OK**.

3. On the **Oneway Analysis of Total Oct By State** hot spot make these choices:

    a. Select **Display Options** and deselect **Grand Mean**.

    b. Select **Display Options** and deselect **X Axis Proportional**.

    c. Select **Display Options** and select **Box Plots**.

    d. Right-click in the box plot graph area and select **Line Width Scale →2.0**.

Thickening the lines for the box plots by selecting a line width of 2.0 will make them stand out a bit more in the display; you should now see a screen resembling figure 3.7.

Figure 3.7 Distribution of totals by state, October 1861

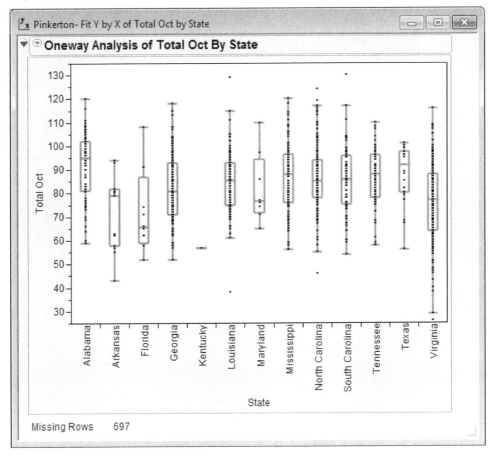

Notice that JMP is displaying individual values (one dot per value) as well as the box plots. Bakker and Gravemeijer (2004) suggest that a display that contains both the individual points and a box plot or histogram as an overlay might aid students in making the transition to thinking about distributions as a "whole." (We can also call attention to the potential small-sample limitations of box plots by referring students to the plots for Arkansas, Florida, and Maryland.) A process called "jittering" will improve on the plot. Jittering adds a small amount of random variation in the display (not the data!). In this case, jittering will present the repeated values for company totals as a set of horizontally placed dots.

1. Click the **Oneway Analysis of Total Oct By State** hot spot again.

2. Select **Display Options → Points Jittered**.

With jittering enabled in figure 3.8 we can acquire a better sense of both the values for company sizes as well as the number of companies from each of the states, something the box plot alone cannot provide. (Recall, however, that if we check **Display Options > X-Axis Proportional**, the differences in sample size will be presented visually by allocating more horizontal box plot space to the states with more companies.)

Figure 3.8 Jittered distribution of totals by state, October 1861

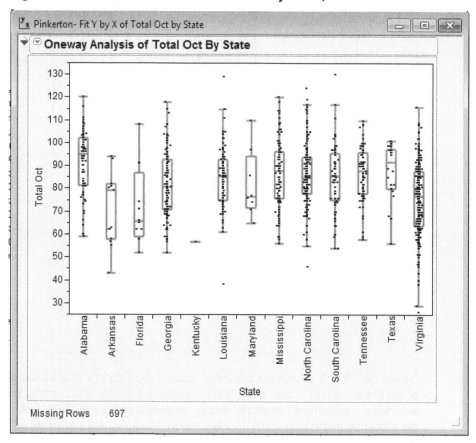

Data One Would Give an Arm For: Blood Pressure

In our first analysis of a large data set I used the Pinkerton data to demonstrate the versatility of JMP as a partner in analyzing large data sets using univariate plots. We compared distributions across time, and across different states. I mentioned in passing some of the thoughts of researchers in statistical education about graphically assisting students as they make a transition from a focus on individual data points to a focus on distributions of data points. My second analysis of a large data set will continue in this vein, mixing new JMP techniques with a discussion of teaching practice.

The techniques of analyzing univariate data can help lay a foundation for bivariate analysis. Cook and Weisberg (1999) and Gould (2004) suggest that looking for trends and patterns in bivariate data can be instructive for students before they actually analyze scatterplots. Garfield and Ben-Zvi (2008) develop an activity based on bivariate data where the explanatory variable has been "binned," that is, the values of the explanatory variable have been transformed into intervals to facilitate looking for trends in data. They believe these plots will help students begin to think about relationships between two variables. Rather than being overwhelmed by a scatterplot, students can examine bivariate data by "looking for a trend as well as scatter from the trend, by focusing on both center and spread for one variable (y) relative to values of a second . . . variable (x)" (p. 290).

Their belief is given experimental support by Noss and colleagues (1999), who performed an experiment designed to assess nurses' understandings of the concepts of average and variation. The nurses had access to statistical software and were asked to consider the relationship between age and blood pressure. The nurses were already familiar with scatterplots, and the scatterplot was their initial choice for a visual representation of the data. Open the BloodPressure file and display the scatterplot of systolic blood pressure versus age, as shown in figure 3.9. This is the scatterplot as it was presented to the nurses. Many of the nurses found it difficult to see a relationship, and some decided there was no relationship between systolic blood pressure and age.

Figure 3.9 Nurses' initial scatterplot

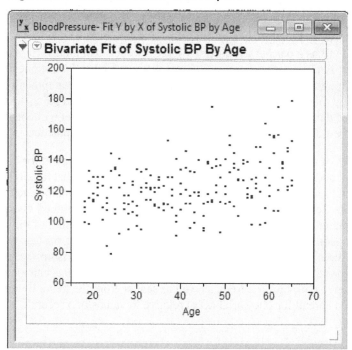

The investigators were puzzled by the nurses' reaction to the scatterplot. The nurses'
prior training and experience should have given them a bias toward a belief in a positive
relation between systolic blood pressure and age; however, the nurses were disappointed
in the perceived weakness of the relationship. The investigators felt part of the problem
was that the large amount of variability in the scatterplot (that is, the relatively large
residuals) may have hindered the nurses' ability to recognize any relationship or trend;
perhaps the nurses believed that "variation and relationship were somehow antithetical"
(p. 42). The investigators prompted the nurses to "slice" the data axis and summarize the
data by finding means within the slices, and by displaying the data as a set of box plots as
shown in figure 3.10.

Figure 3.10 Nurses' box plots

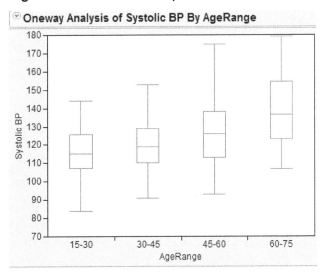

The nurses were initially unfamiliar with box plots, but once they were helped to interpret the plots they were able to see the relationship between age and systolic blood pressure as clearly positive, as well as understand that beyond this simple positive relationship there was substantial variation.

In chapter 5 we will use transformations of numerical data involving elementary functions such as logarithms. Here I would like to use these data to introduce a JMP capability that is often useful when working with categorical data. Our specific goal is to create figure 3.10 using the data in the file BloodPressure. We need to create a variable that expresses the patients' ages as categories by using a method of "recoding" values of the existing ages.

1.   Double-click at the top of the column to the right of the **SystolicBP** variable.

JMP will respond by creating a variable initially named **Column 3**. We will define a new variable, **AgeCategory**. Our plan is to convert ages in the interval [15, 30) to "1," ages in the interval [30, 45) to "2," and so on. After a little head-scratching and inspection of the data values I conjured up the following method: Find the greatest integer (round down) of the age divided by 15. This procedure is shown symbolically in figure 3.11. In addition to creating the variable, we will also need to inform JMP that this variable is categorical rather than numeric. Let us begin this transformation adventure.

Figure 3.11 Greatest integer

$$AgeCategory = \left\lfloor \frac{Age}{15} \right\rfloor$$

2. Double-click **Column 3**.

JMP will respond as shown in figure 3.12.

Figure 3.12 Initial column information

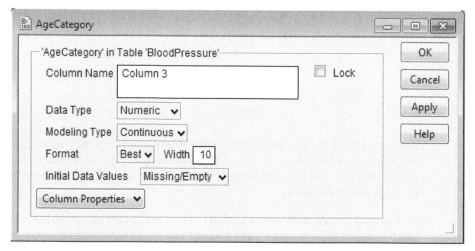

3. Type **AgeCategory** in the **Column Name** field, and select **Character** as the **Data Type**.

4. Select **Column Properties → Formula → Edit Formula**.

JMP responds again, presenting the panel shown in figure 3.13 and informing us that there is no formula (yet). Our task is to make a formula that mirrors our intent.

Figure 3.13 Edit formula panel

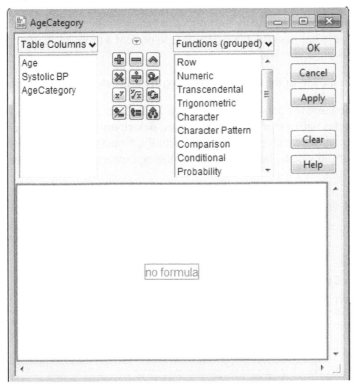

5.   a. Double-click in the **no formula** region.

We will create this variable as a character variable for reasons that will become apparent shortly.

   b. Select **Character > Char, Numeric** → **Floor**, and **Age** → ÷ → **15** → **OK** → **OK**.

As always, check to make sure that the transformation works. Age 19, for example, should appear as an **AgeCategory** of 1, age 30 as a 2, and so on.

Categories 1, 2, 3, and thereafter seem rather arbitrary, possibly more abstract than is comfortable, and almost certainly confusing to those not familiar with the mathematical heritage of the data (that is to say, confusing to a large fraction of the data interpreting public). JMP has a procedure that will allow us to fix this: recoding. The designers of JMP thoughtfully included recoding to handle problems such as inconsistent data entry, typographical errors, and missing data. These are just the sorts of problems that can occur often in a classroom when students gather and enter data (but of course only very rarely when teachers do!). Consider, for example, the potential spelling and capitalization variants of "Rhode Island." Recoding consistently variant data values is easier and less time consuming than individually locating each variant spelling in a large data set. Our use of recoding is not to correct bad data; it is to present a more easily interpreted horizontal axis under our box plots.

6.   Click at the top of the column **AgeCategory** and select **Cols → Recode**.

You should now see the panel shown in figure 3.14. We are going to replace the "numeric" values with a more easily interpreted description; this is why we made the **AgeCategory** variable a character variable initially. Note that JMP will recode, or define new values for old values, but will balk at changing numeric values to character values at the recoding step.

Figure 3.14 Initial recode panel

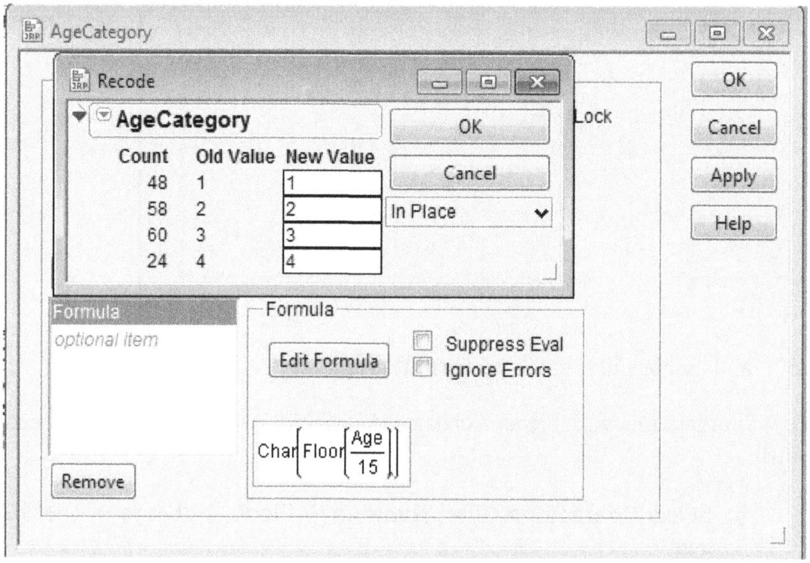

7.   In the field that now has the value **In Place**, select **Formula Column**.

Now replace the **Old Value** of 1 with the **New Value** 15–30. Replace the succeeding **Old Values** with 30–45, 45–60, and 60–75 and click two **OK**s. I am not being too fussy here about the boundary values. If you feel a little squeamish, perhaps something like [15, 30) or 15–<30 would be better, but I am unaware of any decently simple alternative to just explaining in accompanying text what 15–30 means.

8.   Double-click on the new column and change the variable name to **AgeRange**.

You should now see the panel shown in figure 3.15.

Figure 3.15 Variable recoded

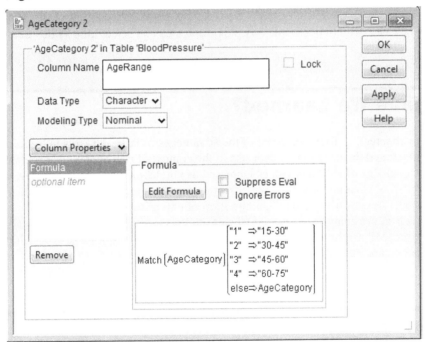

In the **Formula** part of the panel you may notice the **Match** formula. We actually could have used the **Edit Formula** capabilities in JMP to create a formula to do the recoding, but the automatic Recode capability was easier in this case. Now we'll reproduce figure 3.10, mirroring our procedures for the muster roll data. We will implement one additional step we did not take for the muster roll data, so that the data points are not shown in the box plots.

1. Select **Analyze → Fit Y by X → Systolic BP → Y, Response**.

2. Choose **Age Range** for the **X, Factor** and click **OK**.

3. On the **Oneway Analysis of Systolic BP By Age Range** hot spot make these choices:

   a. Select **Display Options** and deselect **Grand Mean**.
   b. Select **Display Options** and deselect **X Axis Proportional**.
   c. Select **Display Options** and select **Box Plots**.
   d. Select **Display Options** and deselect **Points**.
   e. Right-click in the box plot graph area and select **Line Width Scale → 2.0**.

What Have We Learned?

In chapter 3, I discussed some of the advantages of large data sets in teaching and illustrated the use of large data sets in the context of rich contextual problems. Your knowledge of JMP techniques increased as you learned how to define new variables in terms of existing variables. We also considered some thoughts from the literature of statistics education research, in the form of univariate techniques to prepare students for their future analysis of bivariate data. Chapters 4 and 5 will focus on bivariate analysis with JMP, continuing our dual concern with JMP techniques and the teaching of elementary statistics.

References

Bakker, A., & K. Gravemeijer. (2004). Learning to reason about distributions. In D. Ben-Zviand J. Garfield (Eds.), *The challenge of developing statistical literacy, reasoning, and thinking*. Dordrecht, The Netherlands: Kluwer Academic Publishers.

Cook, R. D., & S. Weisberg. (1999). *Applied regression including computing and graphics*. New York: Wiley-Interscience.

Garfield, J., & D. Ben-Zvi. (2008). *Developing students' statistical reasoning: Connecting research and teaching practice*. New York: Springer.

Gould, R. (2004). Variability: one statistician's view. *Statistics Education Research Journal* 3(2): 7–16.

Noss, R., et al. (1999). *Touching epistemologies: Meanings of average and variation in nursing practice. Educational Studies in Mathematics* 40(1): 25–51.

Olsen, C. (2005). Was Pinkerton right? *STATS* 42(1): 24–28.

C h a p t e r 4

The Analysis of Bivariate of Data: Plots and Lines

Introduction

It is difficult to imagine a more important skill than identifying and interpreting associations among variables. In the animal kingdom associating odors or other signals of predators is a survival skill. Human learning builds on associations, such as the association between words and sounds. In the formal empiricism of science as well as in the practical engineering professions, reasoning about the associations induced by experimentation leads to understanding, prediction, and sometimes control of important variables. Understanding the relation between the thickness of beams and their loads allows safe building construction; understanding the relation between dose and response allows identification of safe dosages of anesthesia in the operating room. It is no accident that the formal study of associations among variables plays a prominent role in elementary statistics.

Given our earlier discussions about the difficulties students face when interpreting univariate data, it will come as no surprise that interpreting twice as many variables and their possible relationships is not a simple task. The nature of the difficulty is much more complex than the familiar "correlation implies causation" error. Moritz (2004) and Garfield and Ben-Zvi (2008) and their bibliographies provide extensive discussion of the literature surrounding reasoning about covariation.

The garden-variety scatterplot is used for the graphical display of ordered pairs of continuous data. It is the go-to plot for analyzing questions about relationships between and among variables, and it is not surprising that JMP maximizes the ease of working with scatterplots. In discussing these important plots, I will follow McKnight (1990) and divide reasoning with scatterplots into three parts: (a) observing "facts" as represented by the data points; (b) observing relationships, as represented by the synergistic group of points in a scatterplot; and (c) interpreting and explaining those relationships.

In this chapter, I will demonstrate the capabilities of JMP to facilitate analyses of bivariate data and also consider some of the presentation features of the software. Teachers can use JMP to great effect when teaching the fundamental concepts of data analysis, pointing out interesting features of data. Also, teachers can model good presentation principles for their students with JMP. On the other side of the desk, students can effectively and easily use JMP both for classroom presentations of their projects in real time and in concert with software such as PowerPoint and Word to enhance written reports.

Observing Facts: Pre-Regression

As we know, much of the effort spent interpreting scatterplots is tied up in finding correlations and fitting regression lines to a set of data. In prior statistical studies of univariate data we frequently ask about center, spread, clusters and gaps, and unusual points before we calculate statistics such as means and standard deviations. The data, presented in a graphical form, have a story to tell before the calculation of the statistics. Similarly, bivariate data have a story to tell, irrespective of the correlation and best-fit line. As we do with univariate data, we can ask questions about clusters, gaps, and unusual or unexpected points in the bivariate x-y space. These preliminary questions of the data may actually exhaust the set of reasonable questions, making regression superfluous. For example, suppose we have data summarizing responses to survey questions. The observed pattern of points of mean responses to survey items by two groups, say, male and female, may be interesting and interpretable without the use of regression.

As an example of interpretation without calculation, consider the following data from Jones (1981), a study of young men and women, performed in 1980. Boys ($n = 318$) and girls ($n = 344$) in 28 fifth and sixth grade classrooms in 16 schools in rural towns in 7 counties of North Carolina were asked to indicate their degree of interest in several topics. They rated their level of interest on a seven-point scale (see table 4.1). Each point in the dataset represents a topic of interest, found by averaging the responses of the males and females separately. Figure 4.1(a) shows a scatterplot of the averages of their responses to the items, rendered entirely using JMP tools (**Tools → Annotate** and **Tools → Simple Shape**). I have scaled the horizontal and vertical axes to be the same, ranging from 1.0 to 7.0. What do we see in the scatterplot? With the help of JMP tools we can highlight the fact that the data consist of three clusters and that there are two outlying points. These points are associated with interests in "Motorcycles" and "War." Using **Tools → Line** to add a reference line $y = x$ we can glean some added information, as shown in figure 4.1(b). It appears that both males and females generally agree about what they are interested in, since there is a general southwest to northeast orientation of the points. They seem to agree quantitatively on the points in the low- and high-interest cluster, but the middle cluster garners a higher average interest on the part of girls—the cluster seems to edge in the northerly direction, above the $y = x$ line. Adding the reference line also helps us see something not apparent before; there is a point at which the females exhibit a greater interest than males, a difference as large as the difference in interest between males and females in war. That interest is cooking.

Table 4.1 Interests and gender

Interest	Male	Female
Money	6.49	6.33
Love	6.35	6.47
Life	6.18	6.38
Opposite Sex	6.20	6.00
People	5.61	6.09
Music	5.33	5.96
Sports	6.12	5.23
Peace	5.50	5.84
TV	5.49	5.31
Animals	5.17	5.46
Movies	5.40	5.22

(continued)

Table 4.1 (*continued*)

Interest	Male	Female
Cars	5.70	4.95
Religion	5.07	5.48
Motorcycles	5.70	3.41
Magazines	4.08	4.16
Teachers	3.63	4.64
School	3.30	4.40
Cooking	3.05	4.40
Other Countries	3.58	3.84
Generation Gap	3.49	3.59
War	3.05	1.86
Death	1.98	2.25
Alcohol	1.90	1.74
Cigarettes	1.57	1.71
Drugs	1.59	1.53

Figure 4.1(a) Clusters

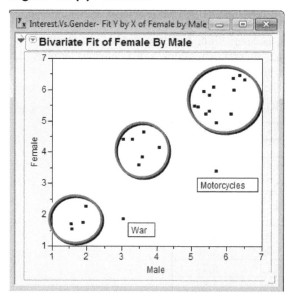

Figure 4.1(b) Outliers

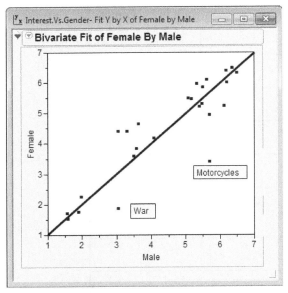

The results of another survey where the pattern of points is of more interest than the regression are presented in table 4.2 and plotted in figure 4.2. These data are taken from a study of parents and adolescents (Feldman and Quatman, 1988). The investigators surveyed 217 parents of male early adolescents and an independent group of adolescent males regarding their views about when adolescents should be able to make decisions autonomously. Each data point represents an individual choice (Child_Mean, Parent_Mean). Once again, JMP allows us to focus attention on an interesting aspect of the graph. In this case, we generally notice an unsurprising result: Parents' "years-of-yes" are typically higher than the adolescents' fond wishes. The added reference line, $y = x$, has also made apparent what we might not have seen at a glance from the numeric data: There are two choices where parent values are lower than adolescent values. Even more surprising is one of the topics: "Go to boy-girl parties at night with friends." The other, "Choose hairstyle even if your parents don't like it," is also interesting. Our point here is, of course, that JMP allows us to focus student (or whatever audience) attention to interesting points, clusters, and trends on a graph. Having said that, it *is* interesting to speculate: Why are these two points below the line? What do these two points reveal about parents and young men? Given the sample size, it is unlikely to be a sampling error. My students, over the years, have consistently come to two conclusions: (1) parents are operating under the mistaken impression that today's boy-girl parties are as sedate as they were in "their day," and (2) with respect to hairstyle, the parents are picking their battles.

Table 4.2 Teen timetable data

	Child	Parent
Choose hairstyle even if your parents don't like it	14.8	14.1
Choose what books, magazines to read	13.2	14.3
Go to boy-girl parties at night with friends	14.8	13.9
Not have to tell parents where you are going	17.2	18.9
Decide how much time to spend on homework	13.0	15.0
Drink coffee	16.0	17.5
Choose alone what clothes to buy	13.7	14.7
Watch as much TV as you want	14.3	17.2
Go out on dates	15.4	16.1
Smoke cigarettes	20.3	20.5
Take a regular part-time job	16.2	16.6
Make own doctor and dentist appointments	17.4	17.9
Go away with friends without any adults	15.8	18.5
Be able to come home at night as late as you want	17.7	19.4
Decide what clothes to wear even if your parents disapprove	15.8	16.0
Go to rock concerts with friends	16.1	17.3
Stay home alone rather than go out with your family	14.5	15.0
Drink beer	18.9	19.3
Watch any TV, movie, or video show you want	15.3	17.4
Spend money (wages or allowance) however you want	13.4	14.1
Stay home alone if you are sick	13.4	14.2

Figure 4.2 Two interesting points

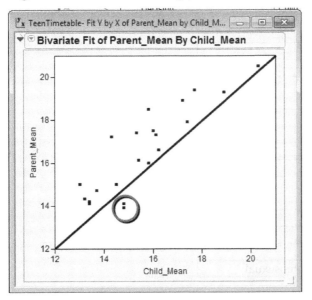

In our last pre-regression analysis of a scatterplot I illustrate a couple of additional JMP features: the capability of comparing scatterplots and of using different "markers" for points in the scatterplots.

These data are from a study of college men and women (Pierce and Kirkpatrick, 1992). Thirty female and twenty-six male residents, recruited from introductory psychology classes at Purdue University, completed a survey of seventy-two "fear items" (see table 4.3). A fear item was defined as something "other" people have reported they fear. The researchers believed that responses to such questions actually tap the respondent's attitudes. Approximately one month after the first survey, a second survey was given that contained 25 items, 14 of which were duplicates from the first survey. Before the students took the second survey they were hooked up to a machine that they were told monitored heart rate, a measure often used in lie detector tests.

Table 4.3 Fear topics

Fearful Topics
Harmless fish
Mice
Rats
Roller coasters
Harmless snakes
People with AIDS
Crawling insects
Death of a loved one
Harmless spiders
Taking written tests
High places on land
Speaking in public
Enclosed spaces
Idea of you being a homosexual

The results for the common items on the two tests for the men and women are shown in figure 4.3.

Figure 4.3 Comparing scatterplots

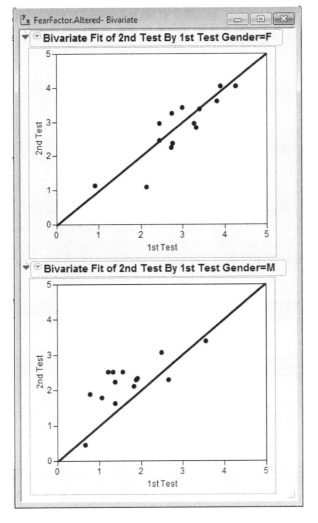

The data are in the JMP file, FearFactor. A peek at the data in this file appears in figure 4.4. Notice that **Gender** is a character type variable. The column on the left, at the business end of the arrow, indicates the row number for the data.

1. Click and drag in this column rows 1 to 28. The columns will then have a "filled in" appearance.

2. Right-click in the column and hover over the box where it says "markers." A list of possible markers for the points will appear.

Figure 4.4 The row number column

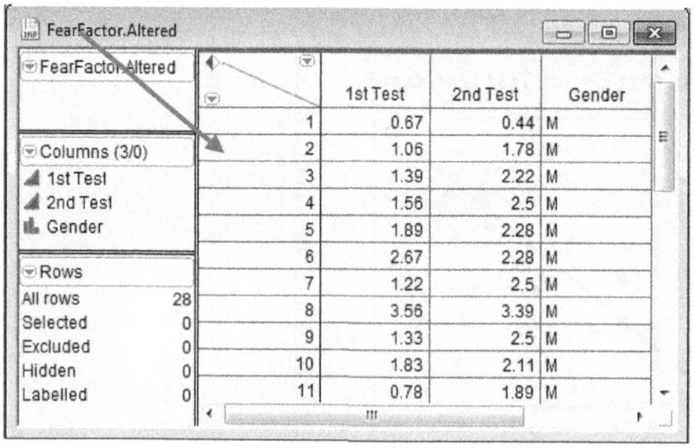

I have chosen as a marker the larger filled circles, but you may choose your favorite from the list. (If your favorite does not appear, feel free to choose **Custom** and make your selection.) It is also possible to have different markers for males and females. To change a marker for, say, males, click and drag in the column rows 1 to 14 and choose a different marker.

3. Select **Select Analyze → Fit Y by X**.

4. Select the variables as shown in figure 4.5. Then choose **OK**.

5. Click each of the 4 scales (2 horizontal and 2 vertical), and set them to a minimum of 0 and maximum of 5, with an increment of 1.

6. Use the **Tool → Line** sequence as before to insert the $y = x$ lines.

Figure 4.5 Set-up for comparing scatterplots

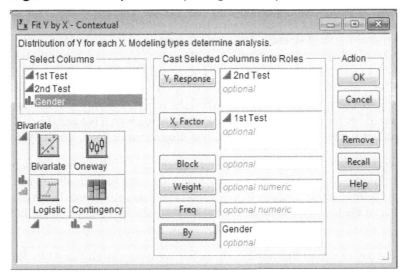

We are now ready to inspect the results. A couple of interesting observations may be noted in these two plots. First, it appears that the female fear scores are clustered at higher values with a couple of possible outliers at the low end ("Harmless fish" and "Idea of you being a homosexual"). The males report generally lower fear scores with two points relatively far away from the herd, one low ("Harmless fish") and one high ("Death of a loved one"). Another observation might focus on the locations of the points relative to the $y = x$ line. Recall that in the second test the individuals were primed to think that the machine might detect deviations from the truth. The points for females scatter more or less about the $y = x$ line, but the points for males tend to be elevated above the line in this second survey. The investigators concluded that the males might have been slightly less than truthful about their fears in the first survey.

Crime Scene Investigation Regression: Murder Most Foul!

We now turn to the topic of linear regression. We shall consider the extensive capabilities of JMP for performing regression, demonstrate how to check the assumptions for regression, and then turn to techniques of transforming data to achieve linearity.

For a regression analysis context we will turn to the field of crime scene investigation (CSI) and reopen a notorious nineteenth-century homicide case, one that from this point in time can only be regarded as a *very* cold case. Readers should not their hopes up; unlike the usual television program fare, we will not solve the case. We will only attempt to shed some light on the old forensic evidence with the benefits of 20/20 hindsight, modern statistics, and science. The crime of interest occurred on August 4, 1892, in Fall River, Massachusetts, where the sun was shining on a pleasant New England day. At about 10:45 a.m., in a two-story house at 92 Second Street, Miss Lizzie Borden helped her father, Andrew Borden, settle comfortably on a sofa for a nap. Shortly before 11:00 she went to a barn behind the house. When she returned twenty minutes later, she discovered her father horribly murdered, slaughtered with a hatchet. Shortly after, the body of her stepmother, Abby Borden, was discovered upstairs, similarly brutally slain.

The police initially believed the murderer had come in off the street. Perhaps it was a grudge killing; Andrew Borden was a banker, and some individuals reported that he had recently refused to loan money to a disgruntled citizen. But, considering the excessive violence of the murders, some homicidal hatchet-wielding maniac might have done the deed.

Within a day or two of the murders the police were convinced that Abby Borden was murdered an hour or two before the father. Such a large time lapse tended to rule out the intruder theory, and suspicion settled on Lizzie as the likely murderer. The strongest argument for the time lapse between the murders came from autopsy reports, specifically the level of digestion of stomach contents. Dr. Edward Wood, professor of chemistry at Harvard Medical School, concluded that Abby Borden must have perished about an hour and a half before Andrew Borden. Lizzie's alibi for her father's murder was that she was in the barn, a sort of storage area for the family, looking for some metal. She could offer no explanation for how the brutal and presumably noisy murder of her stepmother could have occurred an hour and a half earlier when she was in the house.

In 1892 very little was known about gastric emptying times (GET). As late as 1980 evidence on GET, especially the effects of age on GET, was contradictory. In 1984, researchers finally got some hard data. Horowitz and colleagues (1984) brewed up a "hamburger" containing some 40 millirads of radioactive material mixed in with some food and somehow found some volunteers to ingest it. They then measured the time it took to digest 50 percent of the food. Subjects older than sixty-five were recruited from an elderly citizens' club. Our goal in this analysis will be to see what light, if any, these data shed on the forensic case against Lizzie Borden. The data consist of the ages in years, and the GET in minutes, for the subjects in the study.

To follow along with our analysis of the 50 percent emptying time, open the LizzieBorden.jmp file. The data should appear in the JMP table as shown in figure 4.6. Notice that I have again changed the markers to large filled-in circles. You will need to click and drag over the left column to implement the plotting as larger circles in the graphs.

Figure 4.6 Initial GET data

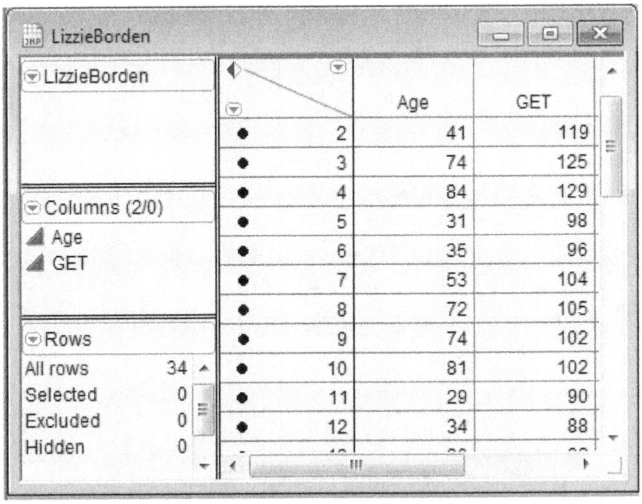

		Age	GET
•	2	41	119
•	3	74	125
•	4	84	129
•	5	31	98
•	6	35	96
•	7	53	104
•	8	72	105
•	9	74	102
•	10	81	102
•	11	29	90
•	12	34	88

LizzieBorden

LizzieBorden

Columns (2/0)
- Age
- GET

Rows

All rows	34
Selected	0
Excluded	0
Hidden	0

The first step in a regression analysis is to create the scatterplot and fit a line to the data. We don't necessarily have any *a priori* reason for believing the relationship between the two variables is linear, but a linear relation is the default simplest relation. Thus, the linear fit is our first option. If the relation turns out to be more complex, the scatterplot and residual plot will perhaps give us clues about the nature of the complexity. (You may need to refer to chapter 1 to review how to get the plot and regression line.) Inspecting figure 4.7, we see that the line is a reasonable fit to the data, although the relation does not appear to be a very "strong" relation. There is a great deal of variability about the best fit line, and r^2 as reported by JMP is only about 0.16.

Figure 4.7 Regression of GET by age

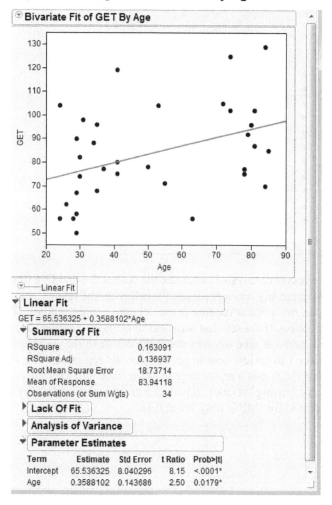

⌄ **Bivariate Fit of GET By Age**

⌄———Linear Fit

⌄ **Linear Fit**

GET = 65.536325 + 0.3588102*Age

⌄ **Summary of Fit**

RSquare	0.163091
RSquare Adj	0.136937
Root Mean Square Error	18.73714
Mean of Response	83.94118
Observations (or Sum Wgts)	34

▸ **Lack Of Fit**

▸ **Analysis of Variance**

⌄ **Parameter Estimates**

Term	Estimate	Std Error	t Ratio	Prob>|t|
Intercept	65.536325	8.040296	8.15	<.0001*
Age	0.3588102	0.143686	2.50	0.0179*

The statistical procedures involved with regression allow us to investigate the relationship between two variables by constructing a "straight line" model of that relationship. Some important assumptions underlie the regression procedures, and these should be checked for credibility before we consider the regression model as adequately justified for inference purposes. The first assumption is, of course, that a straight line is a good representation of the relationship between x and y. If the relation were perfect and deterministic, things would be really simple: $y = \alpha + \beta x$ would perfectly describe the relation. However, in statistics (unlike elementary algebra) our models must exhibit at least some passing resemblance to reality, including all those real-world random events that occur when measuring and sampling in the field. Thus, our model is constructed in terms of what response we expect to see for a given x:

$$E(y|x) = \alpha + \beta x$$

More commonly, we acknowledge explicitly that our response variable, y, has an element of "error" in our final model. So we can make two sorts of assumptions about the relationship between x and y: (a) the average or "expected" behavior is a linear function, and (b) we acknowledge that the value of y for a given x will deviate from expectations in a chance-like way. We will represent this "error" due to chance using the Greek symbol ε, and our final model can be written as:

$$y = \alpha + \beta x + \varepsilon$$

The two sorts of assumptions we make are about the structure of the expectations ("linear") and the distribution of the errors ("normal with zero mean and constant variance"). If our assumptions don't match the real-world situation, there could be serious repercussions for anyone who trusts our analysis and also neglects to verify. Our statistical credibility and fine-tuned sense of statistical responsibility lead us to the right path: Trust if one must, but verify when one can—and one can trust JMP to make the verification very easy to accomplish.

We first consider the assumption that the relation between GET and Age is linear. Our initial look at the scatterplot doesn't generate any suspicion of nonlinearity, but perhaps there is a slight curve in the data that we can't see when our *GET* scale runs from the neighborhood of 50 to the neighborhood of 130. A residual plot will send a much clearer signal.

JMP makes it very easy to create a residual plot, and just at the time it is needed.

1. Select the hot spot **Linear Fit → Plot Residuals**.

Something similar to figure 4.8 should appear just below the parameter estimates. Notice that the residual plot presents the residuals versus the explanatory variable, which makes perfect sense for simple regression. Some statistical programs will default to plotting the residuals versus predicted response variable—*GET* in this case. A residual plot with predicted response variables is certainly appropriate for multiple regression analysis, but for simple regression the residual plot with the explanatory variable seems intuitively clearer. JMP will, of course, perform to your specifications, and if you wish to see the residuals plotted versus the predicted GET scores, that is not a problem. Return to the **Linear Fit** hot spot, and

2. Select **Linear Fit → Save residuals** and then **Linear Fit → Save Predicteds**.

Figure 4.8 GET residuals

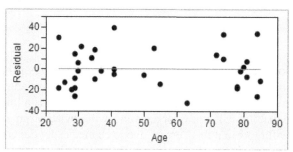

The residuals and the predicted GET values are added to the JMP data table as shown in figure 4.9. JMP gives the residual plot quickly and easily, as you can verify by clicking on **Linear Fit → Plot Residuals**.

Figure 4.9 Residuals and predicted values added

		GET	Residuals GET	Predicted GET
●	2	119	38.7524577	80.2475423
●	3	125	32.9117218	92.0882782
●	4	129	33.32362	95.67638
●	5	98	21.3405596	76.6594404
●	6	96	17.9053188	78.0946812
●	7	104	19.4467356	84.5532644
●	8	105	13.6293421	91.3706579

I will sometimes use an alternative method for students to drive home the point that the regression algorithm produces residuals uncorrelated with the predicted response variable and with a mean of zero. I construct a scatterplot and perform a simple regression for residuals versus predicted response values. For any linear regression, the line for "predicting residuals" must always be residual $= 0x + 0$. I am sure you remember how to do this, but just in case here is my sequence of operations:

3. Select **Analyze → Fit Y by X; Predicted GET → X, Factor; Residuals GET → Y, Response; OK; Bivariate Fit Contextual Popup Menu → Fit Line**.

The results are shown in figure 4.10. Note that students may be initially confused by the presentation of equation of the best fit line (arrow) for these residuals; the best fit line in a residual plot is theoretically $\overline{GETResid} = 0 + 0x$. Usually a gentle reminder about scientific notation with negative exponents and the limitations of binary computer arithmetic will convince them that the 5.53e-14 and the 9.281e-16 are both sufficiently close to zero to allay their fears.

Figure 4.10 GET residuals redux

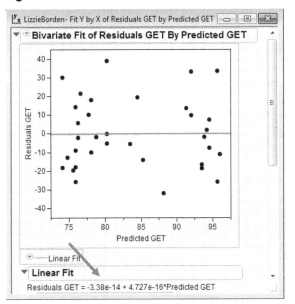

Inspection of our data so far suggests that it is reasonable to proceed as if the average, or "expected" behavior, of *GET* and Age is a linear relation. To determine the quality of the fit to the data, we now need to consider our second set of assumptions: The distribution of errors behaves as normal and behaves randomly. As a reminder, the linear regression procedures are based on assumptions that the distribution of errors behaves in a particular chance-like way:

- For each value of x, the errors have a mean of zero.
- The variance of the errors is a constant across the values of x.
- The errors are independent.
- The errors are normally distributed.

Of course we do not actually see the errors, any more than we know the true population intercept and slope in our model. With real data it is convenient to think of the residuals as a "sample" of the errors. If the errors in our model are misbehaving, that should show up in analysis of the residuals. Thus, it is to these residuals we now turn.

At this point a small technical side note may be in order. Instructors with mathematically more capable students may wish to analyze the residuals using "standardized," or possibly "PRESS," residuals in preparation for multiple regression. I note in passing that these more mathematically intense features are available for simple linear regression in JMP using the **Analyze → Fit Model** keystroke commands.

Recall that we saved our residuals in a separate column, Residuals GET, when we plotted our residuals versus the predicted GET values. The residual plot is not only a good way to check the presumption of linearity, but it is also an excellent plot for checking the homogeneity of variance assumption. I will turn to the Pinkerton data from the Civil War to illustrate this idea. In figures 4.11(a) and 4.11(b), the scatterplot, best-fit line, and residual plot are shown for the December %Sick versus October %Sick for Confederate companies. Notice that with greater values of %Sick in October, there is greater variability in the residuals. This is sometimes referred to as "fanning."

Figure 4.11(a) Scatterplot, %Sick

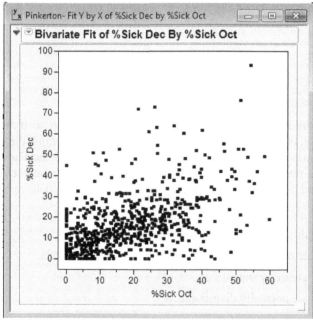

Figure 4.11(b) Residual plot, %Sick

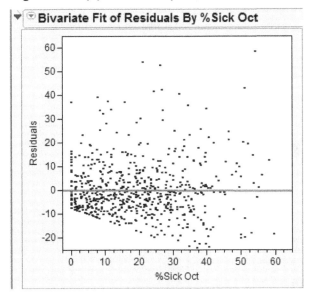

We do not see this pattern in the *GET* versus *Age* plots, which suggests that the presumption of homogeneity of variance is a credible assumption.

The credibility of the assumption of normal errors can be assessed by checking the distribution of residuals. If it appears that the residuals could be a sample from a normal population, the assumption is deemed appropriately made.

1. Select **Analyze → Distribution**.

JMP presents a choice of univariate plots of the residuals for consideration. To get the display in figure 4.12:

2. Select **Distributions → Stack Residuals → Normal Quantile Plot**.

3. Click the box plot and toggle off the **Mean Confid Diamond** and **Shortest Half Bracket**.

Figure 4.12 NQP, GET residuals

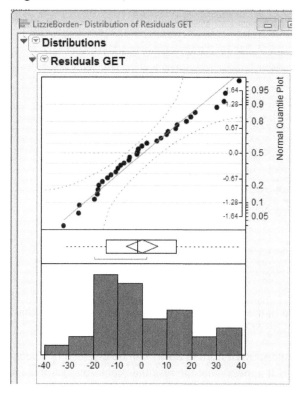

The normal quantile plot appears fairly straight; the boxplot seems symmetric and is bereft of outliers. Even the histogram appears to be cooperating, and to the extent it can with a small dataset, it suggests the reasonableness of the assumption of a normal population.

It appears that we have about as well-behaved data, consistent with a linear relation, as one could hope for. Having said that, which aspect of the regression analysis might shed light on the forensic case against Lizzie Borden? Remember that the autopsies for Mr. and Mrs. Borden showed a large difference in the progress of gastric emptying, with Abby Borden's stomach contents in a far more digested state. It was this that led to the pronouncement of an hour and a half difference in the times of the murders and refocused police attention away from an "unknown intruder" as the likely murderer. But could this difference be attributed to natural variability in digestion rates from human to human? The estimate of that amount of variability is given by the root mean square error (the estimate of the standard deviation of the population errors) in figure 4.7: 18.8 minutes. Is that enough variability to suggest that the murders could have taken place at the same time and just looked like they were committed 90 minutes apart because of natural variation? One authority on the Borden murders (Masterton, 2000) believes this variability to be a significant piece of evidence that tends to exonerate Lizzie Borden. Discretion being the better part of forensic and statistical valor, I shall carefully avoid treading on the guilt or innocence issue.

What Have We Learned?

In chapter 4, I discussed bivariate data plots in contexts with and without regression. I used the annotation capabilities of JMP to demonstrate their use as teaching tools when data are presented and discussed. I also performed a detailed regression analysis and verified the underlying assumptions. With all assumptions holding, it was possible to clearly interpret this regression analysis in a real context. In chapter 5, I present some illustrative examples of more troublesome data.

References

Feldman, S., & T. Quatman. (1988). Factors influencing age expectations for adolescent autonomy: A study of early adolescents and parents. *Journal of Early Adolescence* 8(4).

Garfield, J. B., & D. Ben-Zvi. (2008). *Developing students' statistical reasoning: Connecting research and teaching practice.* New York: Springer Science + Business Media B.V.

Horowitz, M., et al. (1984). Changes in gastric emptying rates with age. *Clinical Science* 67: 213–18.

Jones, R. M. (1981). A cross-sectional study of age and gender in relation to early adolescent interests. *Journal of Early Adolescence* 1(4): 365–72.

Masterton, W. (2000). *Lizzie Didn't Do It!* Boston: Branden Publishing Co.

McKnight, C. C. (1990). Critical evaluation of quantitative arguments. In G. Kulm (Ed.), *Assessing higher order thinking in mathematics.* Washington, DC: American Association for the Advancement of Science.

Moritz, J. (2004). Reasoning about covariation. In D. Ben-Zvi & J. Garfield (Eds), *The challenge of developing statistical literacy, reasoning and thinking* (227–55). The Netherlands: Kluwer Academic Publishers.

Pierce, K. A, & D. R. Kirkpatrick. (1992). Do men lie on surveys? *Behavioral Research Therapy* 30(4): 415–18.

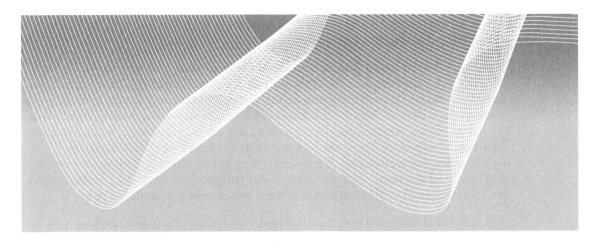

Chapter 5

The Analysis of Bivariate Data—Diagnostics

Introduction

We saw in chapter 4 that JMP is a terrific partner to have when performing regression diagnosis. We breezed through a regression analysis of the gastric emptying time data with ease, generally checking the assumptions for the straight-line model with gratifying results. The data turned out to be decently well behaved, so we could safely operate under the presumption of meeting the assumptions. However, not all data will be benignly distributed, and in this chapter I will demonstrate some of the capabilities of JMP in the context of less well-behaved data than we have previously presented. The skeptical but judicious consideration of the plausibility of the regression assumptions and careful search for troublesome points generally is termed "model adequacy checking" and "diagnosis" of a fitted line (Montgomery, Peck, and Vining, 2006).

Weisberg (1983) refers to *diagnostics* as model criticism and uses the term to refer to statistics rather than graphics. He also separates *diagnostics* from *influence analysis*, his term for a critical focus on individual points. Here I will generally use the term *diagnosis* to mean verifying and checking assumptions associated with fitting a straight line to data. The mathematical statistics underlying model checking is beyond the level of elementary statistics. Montgomery, Peck, and Vining (2006) and Kutner et al. (2005) provide excellent detailed presentations. Fortunately, we can use relatively elementary graphic and numerical diagnostic methods with simple "straight-line" regression. In this chapter we focus on useful JMP techniques applicable to demonstrably nonlinear data.

The analysis of individual points will involve examination of residuals, outliers, and influential points. Simple definitions of *outlier* and *influential* are elusive, and in elementary statistics the identification of outliers and influential points involves a significant amount of subjective judgment, as opposed to strict numerical calculations according to an agreed-upon standard. In the subsequent discussion I will generally follow Montgomery, Peck, and Vining (2006): "Outliers are data points that are not typical of the rest of the data." Hawkins (1980) also captures this idea: "An outlier is an observation that deviates so much from other observations as to arouse suspicion that it was generated by a different mechanism." Montgomery, Peck, and Vining (2006) declare a point influential if "it has a noticeable impact on the model coefficients in that it 'pulls' the regression model in its direction."

Outliers

Outliers are data values that appear to be inconsistent with the general trend of our data. Outliers are problematic because they can arise for any of several reasons, and it may be impossible to identify which of these has produced a particular outlier. One possibility is that outlying values are the result of natural variability in the variable(s) under study; that is, the outliers are not actually inconsistent at all. Variability happens.

A crucially important assumption in any statistical analysis is that variables under study are normally distributed. If the variable is in reality heavy-tailed or skewed, data values that would appear to be outliers under an assumption of normality could instead be perfectly reasonable observations. A second possibility is that an error has occurred in the data collection process, perhaps due to a flawed physical measuring device or human error in recording the data. In some cases it may be possible to determine that an error is a recording error, perhaps a misplaced decimal or a digit reversal at data entry. If data entry records have been judiciously kept, such an error may be corrected by reviewing the data entry process.

Unfortunately, while statistics can be created that may nominate points to be outliers, the statistics alone cannot decide the issue. Identifying a value as a potential outlier and subsequently deciding on a proper course of action (deleting the data value, modifying the value, noting the value in a footnote) is a human decision-making activity, fraught with subjectivity. The best that may be hoped is that the decision will be successfully guided by the data analyst's knowledge, wisdom, and experience.

In a bivariate setting, points can be outliers for perhaps two reasons: (a) they have an unusual x and/or y value, or (b) they have unusual residuals when a regression analysis is performed. As an example, we will consider the data in the JMP file UnusualBears. These data are measurements on twelve female black bears (*Ursus americanus*) from a study by Brodeur et al. (2008). The investigators were interested in the behavior of these bears in Canadian forests. Over the past thirty years increased logging has pushed these creatures northward into less favorable habitat, and they are now confined to an area equal to approximately 60 percent of their historical range. The more northern forests offer colder winters and shorter growing seasons, a situation that would be expected to influence the bears' ecology.

The bears in this study were snared and weighed, their ages estimated, their necks radio-collared, and their ears fitted with tags for identification. (One can only imagine what their mates thought of all this.) Over the course of three years, the researchers gathered information about the bears' home ranges. The home range of an animal—the typical area over which it travels—is generally a function of the quality of food in the area. If the food is scarce, a larger area must be searched for food. It is this search behavior that was of interest to the researchers. The home range data are given for two time periods: summer/fall and spring.

Three scatterplots from the data are shown below in figure 5.1.

Figure 5.1 Outliers

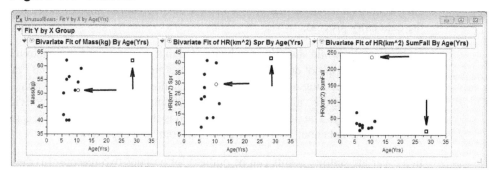

To follow along in JMP for practice, here are the steps to produce the plots using the original file data:

1. Select **Analyze → Fit Y by X → Age(Yrs) → X, Factor**.

2. Select **Mass(kg), HR(km^2) Spr**, and **HR(km^2) SumFall** for the **Y, Response**.

3. Click **OK**.

4. For all three scatterplots change the **Age (Yrs)** scale to run from 0 to 35.

5. For the **Mass(kg) By Age (Yrs)** plot, change the **Mass(kg)** scale to run from 35 to 65, with increments of 5 and 1 tick mark.

6. Click and drag in the row column to darken the markers in the plots.

7. Use the **Tools → Line** feature to add arrows near Bear #5 and Bear #7.

8. Right click on the lines and choose **PointTo** to change the line segments to arrows.

9. Right click on the arrows and choose your favorite colors from the drop-down box.

Note that one could add text boxes to identify the bears of interest; here I chose to highlight the marker capability of JMP. Inspection of the plots reveals that Bear #5 (the square in the plot) is older than the rest (a possible "age outlier"), but she is not particularly different in either mass or home range. Bear #7 (the open circle) is rather pedestrian in age, mass, and spring home range, but she has a large summer/fall home range (a possible "home range" outlier). Inspection of the stem and leaf plots in figure 5.2 will give you a good idea of how far these bears' errant values of age and summer/fall home range fall from their colleagues.

Figure 5.2 Age ranges

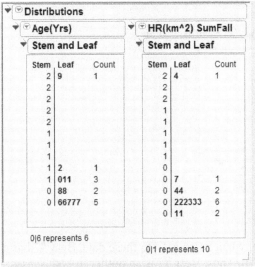

The typical longevity of these creatures is known to be twenty to thirty-five years, so Bear #5 is clearly getting long in the tooth. However, the statistically outlying summer/fall home range of Bear #7 is more difficult to assess. Is Bear #7 a very special bear, or is the outlier attempting to signal that more bears are out there with very large summer/fall home ranges? The researchers did not mention any malfunction of their measuring devices, so that does not appear to be the reason for the large home range. We (and the researchers) are left with two possible explanations for the outliers: They may be unlikely but possible chance occurrences, or they may be signals that older and/or more wide-ranging bears are out there in the northern forests. The statistics and graphs cannot tell us which, if either, answer is correct.

When we perform a regression analysis it is, of course, possible to have outliers in the sense discussed earlier. In addition, once a straight line is fitted to data we may be presented with an additional sort of outlier: points with unusually large residuals. Figure 5.3(a) shows a scatterplot of the home ranges of summer/fall versus spring. Our friend Bear #7 sticks out like a sore paw from the rest of the data. Bear #5 is at the upper end of the spring home range but is otherwise not distinguished in this plot.

Figure 5.3(a) Large residual?

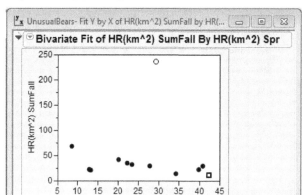

It will come as no surprise that when a line is fit to these data, Bear #7 has a very large residual, as seen in figure 5.3(b). Thus, Bear #7 is an outlier in both senses: well away from the herd and possesses a large residual when a line is fit to the data.

Figure 5.3(b) Large residual!

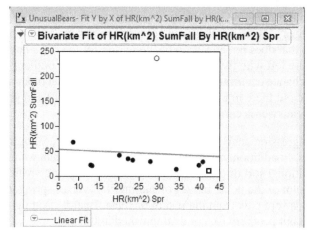

Just how far Bear #7 deviates from the predicted value can be seen with a stem and leaf plot of the residuals as presented in figure 5.4. Recall that JMP will save the residuals in a separate column.

Figure 5.4 HR residuals

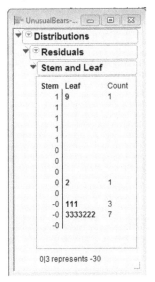

1. Select **Linear Fit → Save residuals**.

We can direct JMP to construct a stem and leaf plot of the residuals using this sequence:

2. Select **Analyze → Distribution**.

Residuals can be standardized in the usual way by subtracting the mean residual (0.0) and dividing by the standard deviation (root mean square error, displayed in the **Summary of Fit** in JMP):

$$\text{standardized residual} = \frac{residual - 0}{RMSE} = \frac{190.537}{64.404} = 2.96$$

The residuals so standardized have a mean of 0.0 and a standard deviation approximately equal to 1.0; a "large" standardized residual potentially indicates an outlier. To standardize all the residuals, create a new column (**StandResids** in my case). After clicking on the top of the newly created **StandResids** column, key in:

3. Select **Column Properties → Formula → Edit Formula**.

We also did this in chapter 3. Then provide the formula as shown in figure 5.5(a). The ordinary and resulting standardized residuals are shown in figure 5.5(b).

Figure 5.5(a) Standardizing the residuals

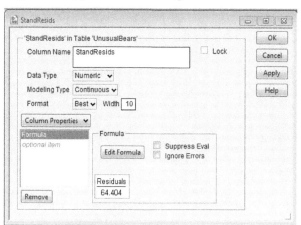

Figure 5.5(b) Comparison

Residuals	StandResids
-20.542473	-0.3189627
-15.441624	-0.2397619
-30.604783	-0.4752
190.536729	2.9584611
-29.686701	-0.460945
-13.307159	-0.2066201
-17.466242	-0.2711981
-13.144001	-0.2040867
-29.162422	-0.4528045
-7.281692	-0.1130627
-29.761828	-0.4621115
15.8621963	0.2462921

The reader will note that I have chosen to demonstrate calculating the standardized residuals with JMP because this standardization process is well known in elementary statistics. Other more advanced methods for scaling residuals exist and are readily available in JMP. Consult Montgomery, Peck, and Vining (2006) for an in-depth discussion of scaling residuals, as well as the **Analyze → Fit Model** sequence in JMP for execution.

I will conclude this discussion of outliers with a teaching observation. My classroom presentation of outliers frequently leads to extended discussions that can be time consuming. As we know, class time is a valuable commodity. Does classroom discussion of outliers possess a sufficiently counterbalancing value add? I believe it does, for two reasons. First, when students go on to analyze data "for real," outliers will appear; their statistics course should prepare them for this eventuality. Second, there is a significant amount of judgment involved in handling outliers; classroom discussion helps cement the notion that differing points of view should be allowed, respected, considered, and debated.

Influential Points

In regression analysis the values of the explanatory variable play a more important role than the values of the response variable. Each of the points has equal weight in determining the intercept of the best-fit line, but the slope is more influenced by values of x that are far removed from $\bar{x}$. Points with unusually low or high x values are said to have high "leverage." Archimedes (287–212 BCE) was quoted as saying, "Give me a lever long enough and a fulcrum on which to place it, and I shall move the world." A modern-day statistician might similarly suggest that given a point with a sufficiently large x-coordinate, he or she could move the slope of the best-fit line. A high-leverage point is not, however, completely sufficient to change the regression world; the power of a single point to change the slope depends not only on the x-value, but also on the y-value of the point in question.

As an example, we consider data from Jeanne and Nordheim (1996). These investigators studied the behavior of *Polybia occidentalis*, a social wasp found in Costa Rica. These data are found in the JMP file UnusualWasps. Jeanne and Nordheim were interested in the forces that determine the swarm size of these social insects, in particular the relationship between swarm size and productivity of the workers. After marking a few adults for future identification, they dismantled their existing nests, forcing active colonies to desert their locations and build new nests. The newly formed nests were left undisturbed for twenty-five days, at which time the researchers collected the individuals from each new colony. They measured different characteristics of the nests, including numbers of queens and workers, and various weights. During nest development, this species of wasp initially has a high number of queens, a number that is eventually reduced to a few or to one. The researchers used weights of the brood and nest as measures of the amount of work performed during new nest construction.

In figures 5.6(a) and 5.6(b) the results of two regressions on number of workers versus number of adults are shown. In figure 5.6(a) the point (1562, 1547) is included in the regression, but in figure 5.6(b) the point is excluded from the regression fit but is plotted. I should note in passing that the incredibly good fit ($r^2 = 0.999$) results from the nature of the variables more than anything else. The adults in a nest consist of queens and workers, and there are many more workers than queens. The target of our focus with these data is the notion that while the point at (1562, 1547) is a point with high leverage (since its x value is unusual), it has little influence on the regression statistics because of its placement very near the regression line.

Figure 5.6(a) Point included

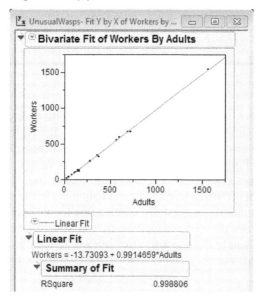

Figure 5.6(b) Point excluded

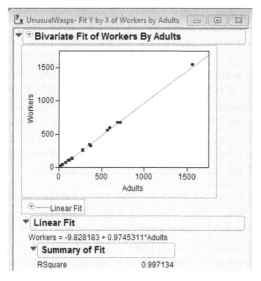

In figures 5.6(c) and 5.6(d) the results of two regressions on the total nest weight versus number of adults are shown. With these data we note that the point at (1562, 23.49) is a point with high leverage (its *x* value is unusually large in this data set), and in addition it has great influence on the regression statistics. The estimate of the slope increases by 68 percent, and r^2 by 24 percent, when the point is deleted. The point is deleted from the analysis but is still plotted by JMP.

Figure 5.6(c) Point included

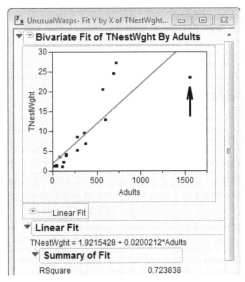

Figure 5.6(d) Point excluded

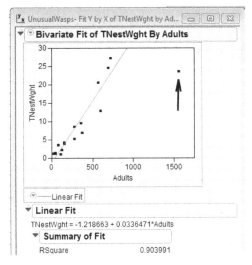

Before moving on to address transformations of data, I will reiterate that it is not a desirable situation if the straight-line model coefficients and/or other properties are greatly affected by a single point or a small group of points. If the influential points are "bad" points, the result of a mis-measure or a nest not representative of the population, they should be corrected or eliminated. On the other hand, if they actually are "good" points the model may have to be re-evaluated.

Diagnosing Linearity: Transformations

It sometimes happens that linear regression is plagued not by isolated outliers or influential points, but by whole sets of points that refuse to array themselves in a straight line. This can happen for a variety of reasons, the most common one being that the relation between two variables is not quite as straightforward as we might initially believe. In some elementary cases of bivariate analysis we can use linear regression in the face of a more complicated relation using the technique of transformation of variables. In elementary statistics, the "logarithmic" transformation is quite commonly used. The logarithmic transformation involves replacing the value of a variable, x, with a "transformed" variable, $\log(x)$ or $\ln(x)$, in the regression analysis. Three regression situations where log transformations turn out to be helpful include:

- y is reasonably modeled by a logarithmic function of x: $y = \alpha + \beta \ln(x)$

- y is reasonably modeled by an exponential function of x: $y = \alpha e^{\beta x}$

- y is reasonably modeled by a "power" function of x: $y = \alpha x^{\beta}$

To illustrate how JMP generally handles transformations we will specifically consider data from each of these situations.

The Log Model: $y = \alpha + \beta \ln(x)$

Opfer and Siegler (2007) reported on an aspect of the development of the numerical estimation skills of children, ages 5 to 10 years. Educators believe that estimation skills are important in the development of mathematical capabilities (NCTM, 1980, 1989, 2000). Numerical estimation involves estimating distance, amount of money, counts of discrete objects—and the locations of numbers on number lines. Studies have indicated that individuals appear to become skillful at different types of these estimation tasks at about the same age, leading to speculation that numerical estimation is a single category of cognitive development. An example of the estimation task presented to the children by Opfer and Siegler is shown in figure 5.7.

Figure 5.7 The estimation task (slightly modified from the original)

0   1000

In the interval above, with 0 and 1000 as endpoints, where would you put 35?

The children were given a set of these numbers and asked to put a mark on the number line that would correspond to the given numbers. The investigators found that as children develop, numerical magnitudes appear to be initially represented logarithmically. In other words, children tend to represent the difference between $1 and $100 as larger than the difference between $901 and $1000. We note here that this actually does make a certain amount of sense—the difference between two and three pieces of chocolate (a 50 percent increase) might seem a lot greater than the difference between ninety-two and ninety-three pieces (a 1 percent increase). Perhaps this logarithmic estimation is firm-wired into our brains at birth and replaced by a more realistic understanding with experience. In any case, chocolate-satisfied or not, children gradually "grow into" a linear representation.

We will use the researchers' data for second graders (Opfer, n.d.) to illustrate a couple of JMP capabilities: (a) fitting nonlinear regression functions, and (b) displaying more than one model fit to data on a single plot. The data are in the JMP file NumberLine.

The initial data should appear as shown in figure 5.8. As explained in chapter 4, I have changed the markers to be large dots in order to make them easier to see in the plots. The **Actual** column contains the true location of the numbers on the number line. The **2nd** column contains the median value for Opfer and Siegler's second-grade subjects' responses to the "Actual" numbers being presented.

Figure 5.8 Opfer's data

		Actual	2nd	4th	6th	Adult
●	1	3	55	17	7	5
●	2	5	59	51	9	5
●	3	7	123	55	21	9
●	4	19	343	255	77	17
●	5	26	393	160	79	29
●	6	72	325	388	199	87
●	7	87	475	277	183	71
●	8	231	577	559	225	243
●	9	391	627	483	335	351
●	10	781	681	751	721	757
●	11	811	673	785	763	799

1. Select **Analyze → Fit Y by X**.

2. Select **Actual → X, Factor**.

3. Select **2nd > Y, Response**.

The first hint of difficulty when analyzing bivariate data usually appears when we try to fit the straight line to the data. The result of this initial fit is shown in figure 5.9. The general trend of the points suggests a square root function or a logarithmic function is more appropriate than a straight-line function. We'll try to fit both functions.

Figure 5.9 Second grade versus actual

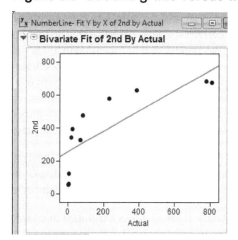

4. Right click on the **Linear Fit** hot spot and choose **Remove Fit** to return to the scatterplot.

5. Click the **Bivariate Fit of 2nd By Actual** hot spot and choose **Fit Special**. The **Specify Transformation or Constraint** options will appear as in figure 5.10. Choose **Square Root sqrt(x)**.

6. Repeat step 4, except choose **Natural Logarithm: log(x)**.

Figure 5.10 Choosing transformations

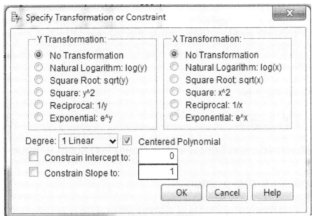

With both fits displayed we can compare the results of the two different transformations as shown in figure 5.11. You can experiment with different colors for the lines by right-clicking on the graph and choosing **Customize**. The fits will have different colored lines, and in practice you can tell which is which by consulting the colors of the line segments shown at the bottom of the displayed fits. It appears that the logarithmic fit is the better fit to the data, consistent with the reasoning of Opfer and Siegler. JMP calculates the coefficients of determination (r^2) for the logarithmic and square root fit for second graders to be 0.95 and 0.80, respectively, suggesting the logarithmic fit is the better one from a raw numerical standpoint.

Figure 5.11 Ln and square root fit

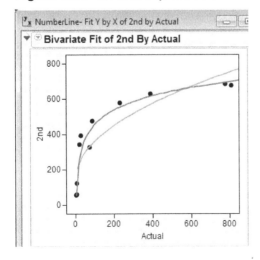

Using the logarithmic model, I will offer an opportunity to practice using the techniques just discussed. In addition to data on second graders, Opfer and Siegler gathered data on the performance of fourth and sixth graders. They hypothesized that as the children grew from second to fourth to sixth grade a logarithmic model would be less appropriate and a straight-line model would better reflect the children's responses. The data for the three grades are in the file NumberLine.jmp.

1. Select **Analyze → Fit Y by X**.

2. Select **Actual → X, Factor**.

3. Select each of **2nd**, **4th**, and **6th** for **Y, Response** as shown in figure 5.12.

Figure 5.12 Choosing three grades

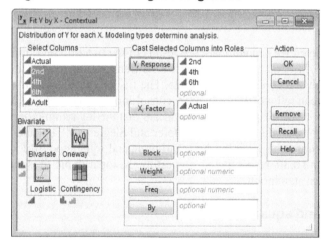

Change the horizontal and vertical scales on each of the graphs to be consistent with those shown in figure 5.13 and fit both the straight-line model and the logarithmic model in each graph.

Figure 5.13 The three grades fitted

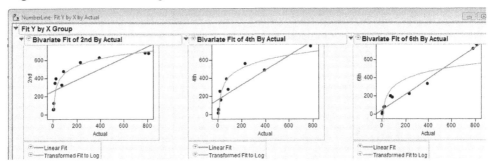

The increasing estimation accuracy as children progress through the grades leads us to believe that older children have a more developed understanding of an interval scale. The changes in coefficients of determination for the two sets of fit, shown in table 5.1, are also consistent with a transition in estimation skills that shifts from one better modeled by a logarithmic model to one better modeled by a straight-line model.

Table 5.1 Coefficients of determination for linear and logarithmic models

Grade	Linear r^2	Log r^2
2nd	0.63	0.95
4th	0.82	0.93
6th	0.97	0.78

Interestingly, the residual plot of the sixth-grade data shown in figure 5.14 provides additional insight about the increase in understanding that is not readily apparent in the scatterplot of the data. The residuals are more varied, with larger residuals for **Actual** values near zero; this suggests that estimation skills may not be fully developed even in grade six. The investigators also ran trials with adults, providing an opportunity to check this conjecture.

Figure 5.14 Residual plot, sixth grade

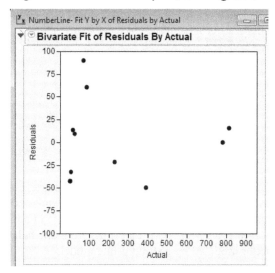

The scatterplot and residual plots, 5.15(a) and 5.15(b), are consistent with the idea that eventually adults develop a "linear representation" of estimation skills. The coefficient of determination for the straight-line fit for adults is a sizzling 0.998, and the residuals appear to have settled down to relatively homogeneous bliss across the spectrum of the **Actual** values. (If anything, the residuals for **Actual** values near zero seem to be smallest for **Actual** values near zero.)

Figure 5.15(a) Regression, adult versus actual

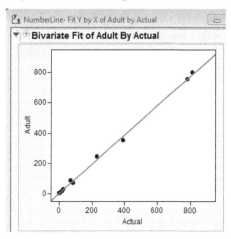

Figure 5.15(b) Residual plot

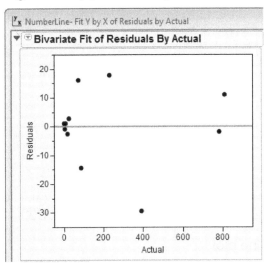

The Log Model Redux: $\alpha + \beta \ln(x)$

Our discussion of transformations for the $\alpha + \beta \ln(x)$ model has been guided mostly by the context of the problem. Opfer and Siegler's analysis of the development of estimation skills on the number line involved children's progression through time from a logarithmic representation to a linear representation of the responses to the number line problem. In figure 5.13 we could see this growth "unfolding" from grade two through adulthood.

In elementary statistics classes it is common to take a slightly different technology tack with transformations as students extend their regression skills beyond the straight-line model, $y = \alpha + \beta x$. That tack is to algebraically force the nonlinear models into a more familiar "linear" form and treat the models as straight lines. The general idea is that explicit use of transformations helps students understand what is going on "under the hood." The transforming process begins with the explicit definition of a transformation of a variable and ends with the recognition that the transformation allows us to use already existing techniques via a newly created straight-line model.

$y = \alpha + \beta \ln(x)$. Define a new variable, x':

Let $x' = \ln(x)$, with the result that:

$y = \alpha + \beta x'$.

This resulting model is then treated as a straight-line model for purposes of analysis. Using the data from Opfer and Siegler's grade two, we created a new, transformed variable, LnActual. The results are shown in figure 5.16.

Figure 5.16 Transformed data

Actual	2nd	LnActual
3	55	1.09861229
5	59	1.60943791
7	123	1.94591015
19	343	2.94443898
26	393	3.25809654
72	325	4.27666612
87	475	4.46590812
231	577	5.44241771
391	627	5.96870756
781	681	6.66057515
811	673	6.69826805

Then we fit a straight-line model to the data, as shown in figure 5.17(a). The logarithmic model fitted previously is also presented for comparison in figure 5.17(b).

Figure 5.17(a) Straight-line plot

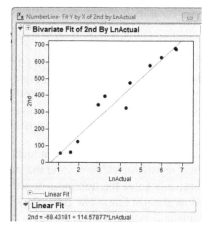

Figure 5.17(b) Ln plot

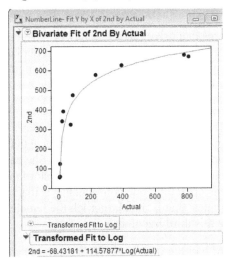

Notice that the equations of the best-fit functions are the same; the *x*-axis scale and the shape of the graph have been changed, essentially because of the new "scale" introduced by the transformation. Where before the **Actual** numbers graced the *x*-axis, we now have natural logs of **Actual** (see figure 5.18).

Figure 5.18 Linear fits, grades versus LnActual

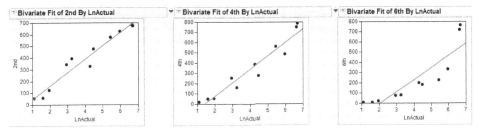

The Exponential Model: $y = \alpha e^{\beta x}$

Another model that is amenable to using logarithmic transformations is the exponential model. Since this model very frequently appears in biological and ecological studies, I will make use of a context taken from those fields: the mating of frogs.

Mating systems in the variety of species in the animal kingdom are incredibly varied (Alcock, 2002). One common mating system is known as *lek polygyny*. A *lek* is a cluster of males gathered in a relatively small area to exhibit courtship displays. The display area itself has no apparent particular value to the females, and their subsequent appearance is for the sole purpose of choosing a mate.

Three major hypotheses about the forces that drive this lek behavior have been proposed. One, the "hotspot" hypothesis, is that males position themselves at particular points (hotspots) along routes of females' typical travel. Two, the "hotshot" hypothesis, is that less attractive males gather with more attractive males (hotshots) to have a chance to be seen or perhaps to intercept females heading for more attractive males.

The third hypothesis is the "female preference" hypothesis. The theory here is that females will prefer larger leks over smaller leks, and it is this preference that drives the males to gather. From the female perspective a larger lek is efficient: She can inspect a greater number of males in less time. In addition, she has a smaller risk of predation in a larger gathering. If female preference is operating, one consequence is that the larger leks should attract proportionally more females; that is, in a larger lek the female to male ratio should be greater. We will investigate this third hypothesis.

Murphy (2002) studied lek behavior of barking tree frogs (*Hyla gratiosa*) and presented three years' worth of data. Data from one of those years, 1987, is presented in the JMP file FrogLeks. (Sorry, couldn't resist the pun.) The data consist of the numbers of males and the numbers of females observed in two ponds over many nights. I fit a simple linear model, $y = \alpha + \beta x$, to the data with result shown in figures 5.19(a) and 5.19(b).

Figure 5.19(a) Fit of female versus male counts

Figure 5.19(b) Residual plot

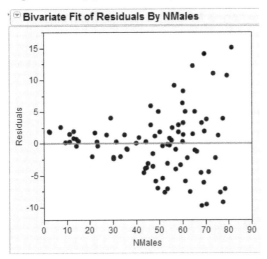

It is clear from the plot that the homogeneity of errors assumption is violated for these data, and the simple linear model will not be appropriate in an inference context. Although we do not necessarily see any obvious curvature in the data, the larger spread of residuals for larger numbers of males in a lek is consistent with an exponential relation, $y = \alpha e^{\beta x}$. We can linearize this relation using the following algebraic steps:

$$y = \alpha e^{\beta x}$$
$$\ln(y) = \ln(\alpha e^{\beta x})$$
$$\ln(y) = \ln \alpha + \beta x$$

This result suggests the transformation, $y' = \ln(y)$. Recall in chapter 3 that we created a new variable by entering a formula that defined a function of existing variables. We will perform this procedure again.

1. Click **Cols → New Column** to bring up the **New Column** panel.

We want to create a new variable, the natural log of the number of females, but we notice that there is a slight problem: Some data points have 0 females, a difficulty for log functions! Our workaround in this situation is to create a variable, $y' = \ln(y+1)$. This has the effect of translating the graph vertically from $y' = \ln(y)$, but the shape of the plot will be unaltered.

2. In the **Column Name** field of the **New Column** panel, type: Ln(NFemales+1).

3. Click **Column Properties → Formula → Edit Formula**.

4. Click **Transcendental → Log → NFemales → + → 1** and close the windows to get back to the JMP data table.

Check to make sure we are agreeing with the formula. The first few lines should appear as in figure 5.20.

Figure 5.20 Check for agreement

	NMales	NFemales	Ln(NFemales+1)
1	2	0	0
2	9	0	0
3	11	1	0.43623677
4	14	1	0.4934887
5	7	2	1.06954816
6	11	2	1.00495006
7	13	3	1.31758047
8	13	2	0.93588909
9	14	2	1.00495006
10	15	2	0.97101563

Now find the best fit line, $\widehat{y'} = a + bx$ and plot the residuals. You should see something like that shown in figure 5.21(a).

Figure 5.21(a) NFemales transformed

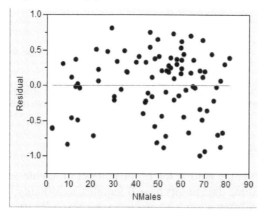

The logarithmic transformation has resulted in residuals that appear to be homogeneous across the number of males in the lek. The slope of the best fit line is positive but not very large, suggesting that the increase in number of females per unit change in the log of the number of males is modest. Perhaps a different perspective on the data may be instructive. We might have reasoned that if the female preference hypothesis is correct, and the number of females in a lek increases at a faster rate than the number of males, then the female-to-male ratio should increase with an increase in the number of males. Let's follow this alternate path. Close any open windows and return to the JMP data table.

1. In the **Column Name** field of the **New Column** panel, type: F/M Ratio.

2. Click **Column Properties → Formula → Edit Formula**.

3. Click **NFemales → ÷ → NMales**.

Once again find the best fit line, this time for $\overline{F / MRatio} = a + b(NMales)$ and plot the residuals. Your results should agree with those shown in figure 5.21(b).

Figure 5.21(b) Ratio transformation

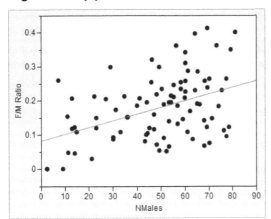

The two analyses are in agreement; the number of females is increasing with the number of males in the lek, consistent with the female preference hypothesis. We should point out that the female preference hypothesis is a causal theory; that is, it is the larger number of males that "causes" the females to prefer joining the lek. However, we have only demonstrated a positive relation with our analysis. In the interest of full disclosure, it should be said that Murphy's (2002) more advanced multivariate analysis in fact casts some doubt on the female preference hypothesis, at least for these creatures.

Our analyses of these data have demonstrated the ease with which data analysts can transform data using JMP, as well as create new variables on the fly, through an intuitive and simple process. Recall that the idea of creating a variable for the ratio of Females to Males did not actually occur to us until after we had performed our initial transformation and regression, and took a new analytical tactical approach. This illustrates a major strength of JMP: JMP responds to new data analysis ideas as they present themselves, helping the analyst explore the data as the results of the unfolding analysis dictate.

The Power Model: $y = \alpha x^{\beta}$

To illustrate the third common model of data that is amenable to the log transformation, the "power" model, we will return to the data from Jeanne and Nordheim (1996) in the file UnusualWasps. Recall that these investigators studied the behavior of *Polybia occidentalis*, a social wasp found in Costa Rica. We can also use these data to explore the use of transformations to minimize the influence of outlying points on the regression. We are interested in the number of cells in these creatures' hives. When studying wasp productivity it would be very time consuming to disassemble the hive and count the number of cells, and we hope to find an easier alternative. If the major component of a nest is the collection of cells with only a small amount of dirt or other incidental material brought in, might we be able to simply weigh the nest and use a regression equation to predict the number of cells? Figure 5.22(a) presents a regression of Cells (number of cells) on Total Nest Weight, and Figure 5.22(b) shows its associated residual plot.

Figure 5.22(a) Unusual wasps regression

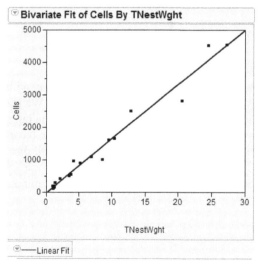

Figure 5.22(b) Unusual wasps plot

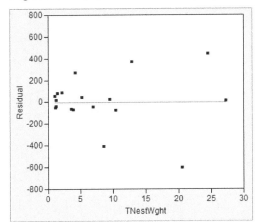

To follow along:

1. Select **Fit Y by X** to get the regression line.

2. Hide the **Analysis of Variance** and **Parameter Estimates**.

3. Click the **Linear Fit context arrow** to **Plot residuals**.

In the Summary of Fit we see that the RSquare is 0.97, indicating apparent success in fitting a straight line to data. However, we also notice that those two nests with Total Nest Weights above 20 are a bit unnervingly far away from the mean Total Nest Weight; although they are very close to the best fit line, they may be exerting significantly more influence than the points closer to (0, 0). Furthermore, the residual plot suggests a difference between the size of the residuals for nests with Total Nest Weights below 10 and those nests with a Total Nest Weight of over 10. It is not difficult to isolate the problem; we see in figure 5.23 that the distributions of both number of cells and the total nest weight are skewed.

Figure 5.23 Indications of skew

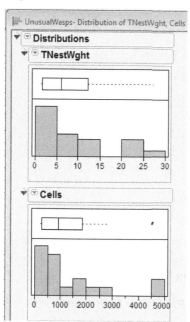

The standard fix for this situation is to implement a logarithmic transformation of both variables and perform the regression analysis on the logarithms of the variables.

We can linearize the power relation as per the following algebra:

$$y = \alpha x^{\beta}$$
$$\ln(y) = \ln(\alpha x^{\beta})$$
$$\ln(y) = \ln \alpha + \beta \ln(x)$$

This suggests the transformations, $y' = \ln(y)$ and $x' = \ln(x)$.

1. Click **Column** → **New Column** to bring up the **New Column** panel.

We would like to create two new variables, the natural log of the number of cells, and the natural log of the total nest weight.

2. In the **Column Name** field of the **New Column** panel, type: LnCells.

3. Click **Column Properties** → **Formula** → **Edit Formula**.

4. Select **Transcendental** → **Log** → **Cells** → **OK** → **OK**, and close the windows to get back to the JMP data table.

5. In the **Column Name** field of the **New Column** panel, type: LogTNestWght. Then click **Column Properties** → **Formula** → **Edit Formula** → **Transcendental** → **Log** → **NestWght** → **OK** → **OK**, and close the windows to get back to the JMP data table.

Check Figure 5.24 to make sure we are on the same page.

Figure 5.24 Post-transformed data

LnNCells	LnTNestWeight
7.82244473	2.55412172
5.27299956	0.20212418
7.93951526	3.02105883
6.98749025	1.93007109
6.79234443	1.65307195
4.9698133	0.22474227

Now find the best fit line, $\widehat{\ln(y)} = a + b\ln(x)$ and plot the residuals. You should see something similar to figure 5.25.

Figure 5.25 Power regression

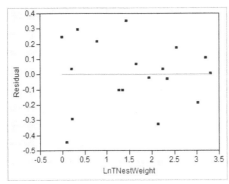

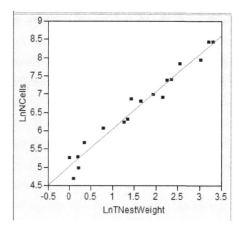

The transformations of both variables using logs has resulted in residuals which appear to be homogeneous across the number of cells. The unequal influence problem has been lessened also, as we can see in figure 5.25; there is less skew in the distribution of both variables.

Figure 5.26 Symmetric data

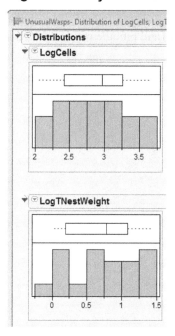

What Have We Learned?

In chapter 5 I focused on some pesky problems that can arise when we use real data to illustrate the analysis of a simple straight-line model. These problems generally turned out to be of two kinds: (a) a few points straying from the general pattern of points, and (b) whole patterns of points doing so. I discussed the identification and display of influential and outlying points for further analysis and for teaching purposes, as well as the facility within JMP with nonlinear relations. We also analyzed three common transformations to achieve linearity using logarithmic transformations in a bivariate setting.

References

Alcock, J. (2002). *Animal behavior: An evolutionary approach* (7th ed). Sunderland, MA: Sinauer Associates.

Brodeur, V., et al. (2008). Habitat selection by black bears in an intensively logged boreal forest. *Canadian Journal of Zoology* 86:1307.

Hawkins, D. (1980). *Identification of outliers*. London: Chapman & Hall.

Jeanne, R. L., & E. V. Nordheim. (1996). Productivity in a social wasp: Per capita output increases with swarm size. *Behavioral Ecology* 7(1):43–48.

Kutner, M. H., et al. (2005). *Applied linear statistical models* (5th ed.). New York: McGraw-Hill.

Montgomery, D. C., E. A. Peck, & G. G. Vining. (2006). *Introduction to linear regression analysis* (4th ed.). Hoboken, NJ: John Wiley & Sons.

Murphy, C. G. (2002). The cause of correlations between nightly numbers of male and female barking treefrogs (*Hyla gratiosa*) attending choruses. *Behavioral Ecology* 14(2):274–81.

NCTM (1980). *An agenda for action: Recommendations for school mathematics of the 1980s*. Reston, VA: National Council of Teachers of Mathematics.

NCTM (1989). *Curriculum and evaluation standards for school mathematics*. Reston, VA: National Council of Teachers of Mathematics.

NCTM (2000). *Principles and standards for school mathematics: Higher standards for our students, higher standards for ourselves*. Washington, DC: National Council of Teachers of Mathematics.

Opfer, J. E. (n.d.). Analyzing the number-line task: A tutorial. Available at: http://www.psy.cmu.edu/~siegler/SiegOpfer03Tut.pdf. Accessed 22 August 2011.

Opfer, J. E., & R. S. Siegler (2007). Representational change and children's numerical estimation. *Cognitive Psychology* 55:169–95.

Weisberg, S. (1983). Some principles for regression diagnostics and influence analysis. *Technometrics* 25(3):240–44.

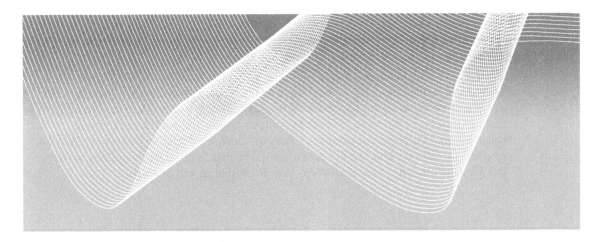

Chapter 6

Inference with Quantitative Data

Introduction

This chapter will explore the general topic of inference with quantitative data by offering a sequence of vignettes for hypothesis tests and confidence intervals. Since some steps in these processes are the same for both hypothesis testing and confidence interval construction, I will try to avoid unnecessary duplication of reading and typing. As you progress through the chapter, notice the natural sequence of selection options as they facilitate the inference procedures—by now a familiar aspect of JMP.

Inference about a Population Mean

Perils of the Anterior Cruciate Ligament (ACL)

Suppose you are walking along and suddenly hear your knee pop, followed closely by your noisy knee giving way, followed by painful swelling. Your body might be trying to tell you that you have just injured your anterior cruciate ligament (ACL). This is not good: You are in for surgery and weeks of rehabilitation. The surgery aims to reconstruct the knee and involves grafting the ACL to the tibia and femur. The grafting is accomplished using what are known as "fixation" nails.

A basic goal in this type of surgery is to prevent the tendon from rupturing again, leading to the question: How much force is needed to detach the fixation nail? That is, how much force can a fixation nail withstand without failing? To address this question, Aydin and colleagues (2004) performed ACL surgery with fifteen human cadaver knees and a particular brand of fixation nail. After the surgery, "tensile loading was applied parallel to the longitudinal axis of the tunnel. . . . Tensile loading at a velocity of 50 mm per minute was applied to the femur. The loading was continued until graft rupture and failure load was recorded."

In other words, they just kept pulling harder on the fixation nail until it failed. We will use their results and the capabilities of JMP to estimate the mean pullout force for this type of fixation nail in ACL surgery by constructing a 95 percent confidence interval. These data are in the JMP file ACLSurgery.

1. Select **Analyze → Distribution → PullOutForce(N) → Y, Columns → OK.**

Before constructing the confidence interval we must assess the credibility of the assumption that the population is approximately normal. Two popular plots used for this purpose are the box plot and the normal probability plot. In chapter 1, we discussed the steps in generating these plots, so we will simply present the results in figure 6.1 (I have suppressed the mean confidence diamond and shortest half bracket). Neither plot gives any indication of skew, so it is safe to regard the presumption of a normal population as credible.

Figure 6.1 Checking assumptions

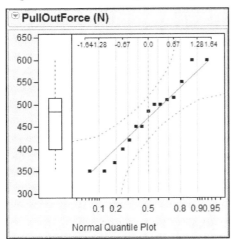

2. Click on the contextual help pop-up triangle for **PullOutForce(N)** and select **Confidence Interval → 0.95**.

The default interval is a two-sided 95 percent confidence interval, and your results should appear as shown in figure 6.2(a). Of course, it is always an option to pick a different confidence level.

Figure 6.2(a) 95 percent CI

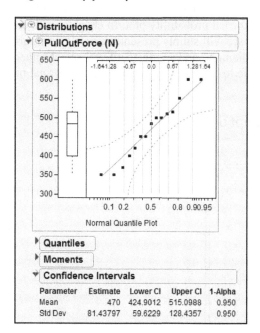

3. Select **Confidence Interval → other** and type 0.80 in the confidence interval field.

As an example, the presentation in JMP of the 80 percent confidence interval for the population mean is in figure 6.2(b).

Figure 6.2(b) 80 percent CI

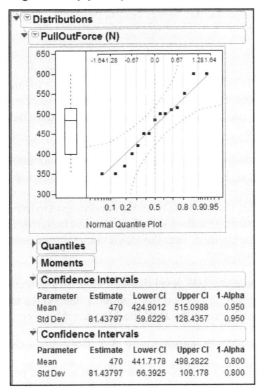

It appears that when we recover from the surgery and rehabilitation, we should avoid any shearing forces on our knees that exceed about 440 newtons.

Magnetic Monarchs?

It is well known that a wide variety of animals use magnetic fields to determine their orientation. The homing pigeon is probably the most familiar example, but migratory birds such as the European robin (*Erithacus rubecula*) and indigo bunting (*Passerina cyanea*) also use the Earth's magnetic field. While the influence of the Earth's magnetic field during migration is well documented, it is not clear in every case how individual species might detect magnetic fields. For detection to occur, the animal must have some sort of magnetic material in its body.

The monarch butterfly (*Danaus plexippus*) is a familiar animal that migrates over tremendous distances. We will use the capabilities of JMP to address the question of whether these creatures might navigate using magnetic fields.

In an exploratory step toward understanding how the monarch butterfly orients itself for travel, Jones and MacFadden (1982), using extremely sensitive cryogenic magnetometers, attempted to determine if monarchs might have some magnetic material in their bodies. They reasoned that if magnetometers detected the presence of such material, the case for monarchs' use of the magnetic field of the Earth to help orient their migration would be strengthened.

Unfortunately a magnetometer creates "background" magnetism, about 200 pico-emus of magnetic intensity. (A pico-emu is 10^{-12} electromagnetic units.) In order to demonstrate that monarchs' bodies contain magnetic material it must be shown that the magnetic intensity of the butterflies exceeds the machine's background magnetic intensity of 200 pico-emus. The researchers prepared a random sample of sixteen butterflies by bathing them in distilled water before placing them near the magnetometer to measure their bodies' background magnetic intensity.

Is there sufficient evidence at the $\alpha = 0.05$ level that the butterflies possess magnetic intensity in excess of 200 pico-emus? Jones and MacFadden's data on the measured magnetic intensity for each butterfly are in the JMP file MagneticMonarch.

1. Select **Analyze → Distribution → MagIntensity → Y, Columns > OK.**

As always, we must check the credibility of the presumption of a normal population. The results are shown in figure 6.3. Neither the box plot nor the normal quantile plot suggests any problems, and we are once again comfortable with the presumption of a normal population.

Figure 6.3 Checking assumptions

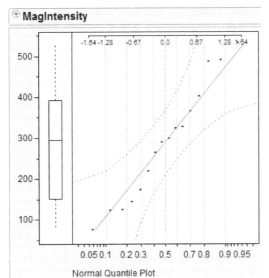

Normal Quantile Plot

2. Click on the contextual help pop-up triangle for **MagIntensity** and select **Test mean**.

3. Enter the hypothesized value, 200, and click **OK**.

My results are shown in figure 6.4. Notice the nice graphic representation in JMP of the sample mean as a value and the P-value as an area. In this case our alternate hypothesis is that $\mu > 200$ and we see a P-value of 0.01135. This is pretty clear evidence of a population mean greater than 200 pico-emus. Apparently the monarchs' bodies have at least the possibility of responding to magnetic fields. Now, here's an added treat from JMP.

Figure 6.4 Testing the mean

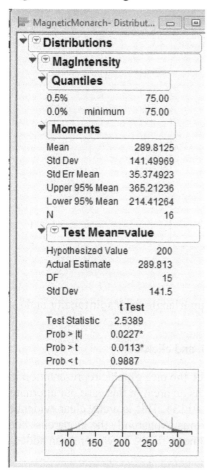

4. Click on the **Test Mean=value** hot spot and select **P-value animation → High Side**.

Drag the handle to adjust the hypothesized mean, and JMP will present a little teaching moment about P-values in figure 6.5(a).

Figure 6.5(a) A JMP P-value script

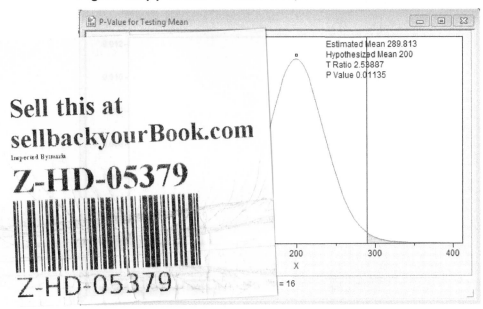

That's not all! As is well known, a perennially difficult topic for students is the concept of power.

5. Click the **Test Mean = Value** hot spot and select **Power animation → High Side**.

The result is shown in Figure 6.5(b). In this panel there are three handles available for manipulation: the estimated mean, the true mean, and the hypothesized mean. You can drag these handles to see how these values affect power in the case of a hypothesis test for a single mean. (Our data has a fixed n, but different sample sizes might be simulated to see the effects of sample size on power.)

Figure 6.5(b) A JMP power script

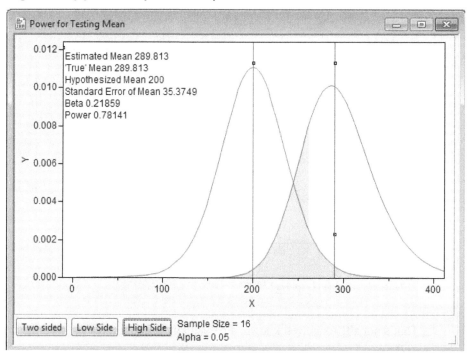

Inference about Means—Paired Data

Monitoring Small Children

We now consider the paired-t procedure. As we know, this is the standard statistical procedure for studies involving two observations that have been taken from the same or similar experimental units. In elementary statistics we see this when we analyze results from a randomized block experiment with two treatments. The statistical analysis is a kind of hybrid. We are interested in the difference between two means but focus on a single mean: the mean of differences. To illustrate the approach taken by JMP to the paired-t, we will call your attention to data from the world of neonatal pediatrics. The data are in the JMP file Finapres.

When doctors evaluate young children (0–4 years old) for possible cardiovascular abnormalities, continuous monitoring of blood pressure is desirable. Historically, blood pressure of adults has been monitored by inserting a catheter with a pressure-sensing mechanism into an artery. This invasive procedure can be dangerous because of the risk of infection and internal bleeding. In adults and children 6–16 years old, a noninvasive method known as "Finapres" (FINger Arterial PRESsure) is used. With Finapres, a small clamping device called a "cuff" is attached to a middle finger on one end and a monitoring device on the other. Some pediatricians have attempted to use Finapres with newborns by wrapping the device around the newborn's wrist but have encountered some theoretical and practical problems. A new, tiny Finapres cuff has been developed and data gathered to compare the indicated blood pressures using the Finapres method and the catheter. (The gold standard is the catheter method.) Andriessen et al. (2008) measured the blood pressure of fifteen children, aged 0–4 years, using the two methods simultaneously. If the cuff were applied to the right middle finger, the catheter would be inserted in the left radial artery (lower arm) and vice versa if the cuff were applied to the left middle finger. The left/right decisions for the cuff were made by random assignment. Their results for diastolic pressure are shown in figure 6.6.

Figure 6.6 Finapres data

	Fin	Cath
1	23	31
2	30	38
3	30	30
4	45	43
5	51	35
6	45	45
7	51	49
8	25	41
9	37	45
10	58	57
11	57	61
12	25	45
13	76	78
14	57	53
15	45	48

The investigators were concerned that the Finapres might be giving a biased estimate of the blood pressure and wanted to estimate the amount of bias. A bias of 0 would indicate agreement on average between the two methods.

We will construct a 95 percent confidence interval for the difference in blood pressure measurements between Finapres and the catheter. Our first step, of course, is to generate these differences.

1. Double-click in the top row of the column to the right of **Cath** to create a new column for the data we will create.

2. Label it **Diff (F – C)**.

That way, we will remember the direction of subtraction without having to return to the data file.

3. Now double-click in the rectangle with the label **Diff (F – C)**.

JMP will respond with the panel shown in figure 6.7.

Figure 6.7 Creating the difference

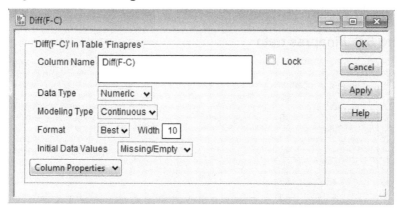

We will need to create a "difference" formula for this variable.

4. Select **Column Properties → Formula → Edit Formula**.

JMP has a rich complement of operations and functions, as we can see in figure 6.8. This might be an opportune time to explore some of the powerful functions listed in the Functions (grouped) panel. When we are finished exploring, we shall return to the land of simple arithmetic.

Figure 6.8 Setting up the difference

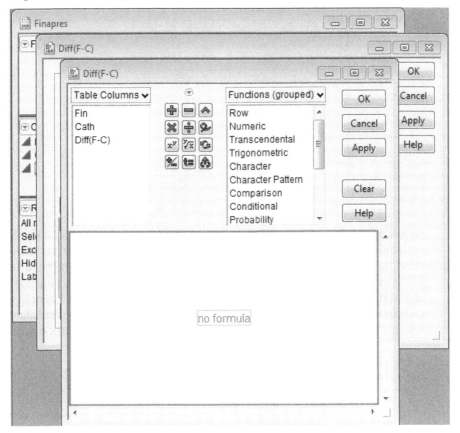

Perform the following sequence of clicks to construct a formula in the "no formula" rectangle, transforming it into a "yes formula" rectangle.

5. **Select Fin → "-" → Cath → OK → OK**.

What you saw in figure 6.6 should now appear with a new column, as shown in figure 6.9. Verify that the calculations have been done correctly, just as you would verify that the data entry is correct.

Figure 6.9 Checking the formula

Finapres				Fin	Cath	Diff(F-C)
			1	23	31	-8
			2	30	38	-8
			3	30	30	0
			4	45	43	2
Columns (3/0)			5	51	35	16
Fin			6	45	45	0
Cath			7	51	49	2
Diff(F-C)			8	25	41	-16
			9	37	45	-8
Rows			10	58	57	1
All rows	15		11	57	61	-4
Selected	0		12	25	45	-20
Excluded	0		13	76	78	-2
Hidden	0		14	57	53	4
Labelled	0		15	45	48	-3

At this point all we need to do is make the confidence interval as we did with the ACL pullout force. Neither the box plot or normal probability plot gives any indication of skew, so the presumption of a normal population is credible. The mean difference, F–C, in the sample is negative: -2.9333. This result suggests the Finapres method underestimates the blood pressure of these youngsters by close to 3 mmHg when compared to the gold-standard catheter. So that you can check your results with mine, the results I get for the Finapres-Catheter difference are shown in figures 6.10 and 6.11.

Figure 6.10 CI for the difference

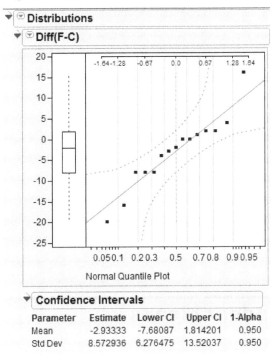

The flip side of the confidence interval is, of course, the hypothesis test. Because so much of the two procedures with paired data is duplicative of our earlier inference for a single mean, we will not present a detailed walk-through. We note in passing that this bias problem might just as easily have been cast as a hypothesis test of "0 bias," with results shown in figure 6.11.

Figure 6.11 Hypothesizing no difference

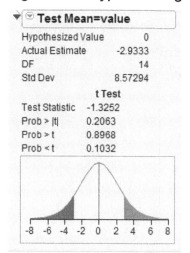

Inference for Independent Means

Lateralization in the Lower Species

Inferring the difference between two population means, given appropriately gathered samples, is one of the most ubiquitous and important techniques in elementary statistics. Thus, it should come as no surprise to the reader that we take up one of the most ubiquitous and important questions of our day: Are snakes left-handed?

It is well known and well documented that higher vertebrates, such as mammals and birds, exhibit lateralized behaviors, commonly described with more than a hint of anthropomorphism as "handedness." Biologists theorize that such behavioral asymmetry is linked to specialized development of brain cells occurring fairly recently (i.e., the time of mammals) in the evolutionary chain. However, some recent studies of lower vertebrates suggest that brain lateralization may have occurred earlier in the chain than previously supposed.

For example, some have suggested laterality occurs in the coiling behavior of snakes. Roth (2003) gathered thirty cottonmouth snakes (*Agkistrodon piscivorus leucostoma*) in his lab and reported on the subsequent observations of their coiling behavior. The cottonmouth is a semiaquatic venomous snake that ranges throughout the southeastern United States. Cottonmouths spend much of their time in a coiled position and are equal-opportunity strikers; both predators and prey fall victim to its venom.

Roth defined a "laterality index" to be the proportion of observed coils that he defined as "left-handed," that is, the snake was curled so that the left side of the body was "pointing in." (Observations of outstretched and random coiling behavior were discarded.) His data reside in the JMP file CottonMouth and also are shown in figure 6.12. In addition to the laterality index, the file contains two categorical variables: age (adult/juvenile) and gender (male/female).

Figure 6.12 Handedness data

LatIndex	AdJuv	Gender
0.582	A	F
0.585	A	F
0.55	A	F
0.554	A	F
0.609	A	F
0.545	A	F
0.544	A	F
0.6	A	F
0.638	A	F
0.656	A	F
0.6	A	F
0.696	A	F
0.424	A	F
0.493	A	F
0.491	A	F
0.512	J	F
0.556	J	F
0.565	J	F
0.417	J	F
0.429	J	F
0.563	A	M
0.556	A	M
0.522	A	M
0.541	A	M
0.395	A	M
0.486	J	M
0.492	J	M
0.475	J	M
0.464	J	M
0.493	J	M

Those whose data analysis experience with technology is limited to a graphing calculator might wonder why the data are not in the familiar "list" format. There are a couple reasons for this. First, at elevated statistical levels problems are commonly formulated as "modeling" problems where the regression format for data works well. Secondly, and more important at the elementary statistical level, this arrangement allows inference for means without having to duplicate data entry of lists. Suppose, as an example, that one wished to test two hypotheses: Adults and juveniles have equal laterality indices, and males and females have equal laterality indices. On the calculator one would have to create two pairs of lists in order to do the tests. With JMP this is not necessary. One simply identifies the desired categorical characteristic as a separate variable, and only a single column of laterality indices is needed irrespective of how many hypotheses are being entertained. Another implication of this "regression layout" will be seen very quickly as we begin making our inferences for independent means.

1. Select **Analyze → Fit Y by X**.

"Fit Y by X" is what we did when we were doing regression back in chapters 4 and 5. While it may seem odd, it makes sense if you think of the laterality index as the response variable and the categorical variables as explanatory variables. That thinking is consistent with the way we think in elementary experimental design. The "treatments" (for example, old drug and new drug) are values of a categorical explanatory variable.

In any case, you should now see the slightly complicated presentation by JMP in figure 6.13.

The reason this window is slightly complicated is that JMP doesn't know yet whether your variables will be quantitative or categorical. It is prepared for any combination of variables, explanatory and response, quantitative and categorical. As soon as I pick the variables to analyze, JMP will determine the correct procedure. In this case our response variable is numeric and our explanatory variable categorical—thus, JMP will select **Oneway Analysis of Variance**.

Figure 6.13 Independent means, first step

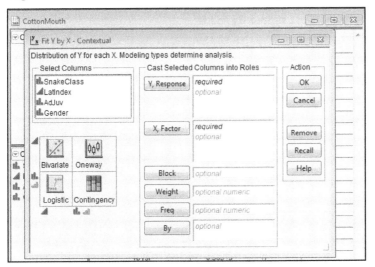

2. Select **LatIndex → Y, Response → AdJuv → X, Factor → OK**.

The appearance of the presentation is similar to what we have seen previously when we analyzed the Pinkerton data. JMP knows only that you will be doing inference for means, but doesn't yet know that you have only two means, so it indicates the forthcoming analysis as the more general statistical technique, Oneway Analysis of Variance, rather than a *t* test as shown in figure 6.14.

Figure 6.14 The one-way ANOVA presentation

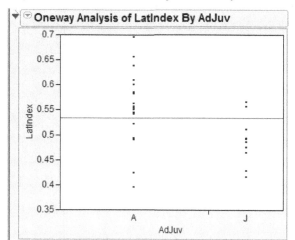

As can be seen in figure 6.14, JMP does know that there are only two values for the explanatory variable, A(dult) and J(uvenile). JMP has placed both values for the explanatory variable (A and J) on the horizontal axis and by default presents dot plots of the values of the laterality index. The dot plots are of limited value in checking the credibility of our presumption of normality, and of course JMP has better options.

3. Click the **Oneway Analysis** hot spot and **Display Options** and (a) deselect **Grand Mean**; (b) deselect **X Axis proportional**; and (c) select **Box Plots**.

4. Click the **Oneway Analysis** hot spot and select **Normal quantile plot**.

You should now see a presentation similar to that shown in figure 6.15. Neither the box plot nor the normal quantile plot gives any indication of problems, so we will proceed with our t-test.

Figure 6.15 Checking the assumptions

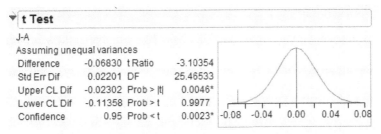

5. Click the **Oneway Analysis → t test**.

The t-test panel is then added to figure 6.15 as shown in figure 6.16. Notice that information relevant to the t-test is presented, as are the bounds for the 95 percent confidence interval. Were a different alpha or confidence level desired, this would be accomplished by selecting **Display Options** and then "Set α level." It appears from the analysis that adult and juvenile cottonmouths have different coiling behavior. (That adults and juveniles differ will come as no surprise to human parents of teenagers.)

Figure 6.16 Testing the hypothesis

▼ **t Test**

J-A
Assuming unequal variances

Difference	-0.06830	t Ratio	-3.10354
Std Err Dif	0.02201	DF	25.46533
Upper CL Dif	-0.02302	Prob > |t|	0.0046*
Lower CL Dif	-0.11358	Prob > t	0.9977
Confidence	0.95	Prob < t	0.0023*

-0.08 -0.04 0.00 0.04 0.08

For extra practice, I suggest that you duplicate these steps and test the hypothesis of equal means for the two snake genders.

Inference for Regression

Sand Scorpion versus Cockroach

The burrowing cockroach (*Anivaga investigata*) and the sand scorpion (*Paruroctonus mesaensis*) share the same desert. Both of these animals burrow in the sand and can detect low velocity surface waves (vibrations in the sand) transmitted by other creatures. The burrowing cockroach and the sand scorpion are both interested in interpreting the vibrations generated by surface waves. Their interests are, however, at cross purposes because the sand scorpion preys on the burrowing cockroach.

The sand scorpion is a nocturnal ambush predator of insects and other scorpions, always hunting from a motionless resting position outside its own burrow. The scorpion detects nearby prey by sensing the small radiating disturbances made by the prey. These disturbances are in the form of small waves, similar to sound waves in the air. The sand scorpion's range, accuracy, and speed in targeting allow it to get to a burrowing cockroach 50 centimeters away in about three rotation-and-translation steps and in three seconds. Speed and accuracy are needed in this case because the cockroach detects scorpion vibrations and will try to burrow deeper in the sand for safety. These data are in the JMP file Scorpions.

When prey appears on the horizon within about 20 centimeters, the scorpion assumes an alert posture and with quick rotations and translations of position grasps and immobilizes the prey for stinging. Within 10 centimeters, prey is almost always captured in one rotation and translation of position by the scorpion. Brownell and van Hemmen (2001) investigated possible biological mechanisms that the scorpion might use, and our data is taken from the control group in their fascinating experiment. If the scorpions are accurately targeting their prey for all angles, it seems reasonable that a regression equation that predicts the response angle (ρ) of the scorpion from the actual angle (α) of the cockroach prey would be $\rho = 0 + 1.0\alpha$. We will unleash JMP to study this issue.

After reviewing chapter 4, we have generated the least squares regression line and residuals. As shown in figure 6.17, the actual angle is measured in terms relative to the scorpion. 0 degrees indicates that the prey is directly in front, 90 degrees indicates that the prey is to the right, and so on.

Figure 6.17 Actual angles

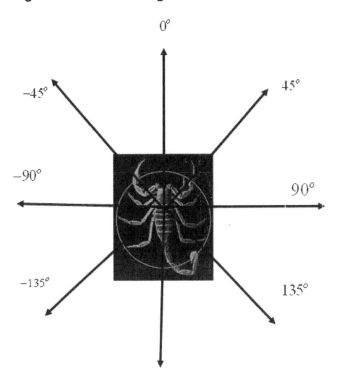

The residuals are shown in figure 6.18, and the regression line is shown in figure 6.19. It appears from a glance at the residual plot that there is a tendency for the scorpions to be less accurate if a cockroach is behind them ($\alpha < -90^\circ$, $\alpha > 90^\circ$). The plot clearly seems to indicate a linear model is appropriate.

Figure 6.18 The sand scorpion residual plot

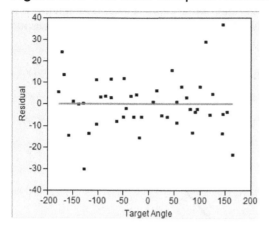

Figure 6.19 Sand scorpion inference

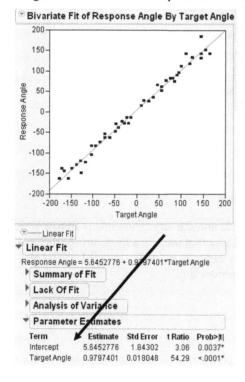

The usual first step when performing inference in a regression context is to perform a Model Utility test, that is, a test for slope = zero. Rejection of the hypothesis indicates that the explanatory variable does indeed help to explain the values of the response variable beyond the customary levels of chance. JMP declares this P-value for the *t*-test to be less than .0001. (To give the real value of the P-value probably "Pico-Ps" would be required.) Notwithstanding the usual first step, it is normally more revealing to look at confidence intervals for the parameter estimates.

1. Select the **Linear Fit** hot spot pop-up triangle and then select **Confid curves fit**.

This command adds "confidence curves" to the fit. These curves are slightly different than confidence intervals, because there are two parameters, the slope and intercept, that go into making the curves.

To find confidence intervals for the individual parameters:

2. Right-click in the **Parameter Estimates** report (in the report, not on the Parameter Estimates title; see arrow) and select **Columns → Lower 95%** and **Columns → Upper 95%**.

This information is added to the parameter estimates as shown in figure 6.20 (see arrow). If you wish, you can select a different confidence level through the **Linear Fit** contextual help pop-up triangle.

Figure 6.20 Confidence intervals for regression

Term	Estimate	Std Error	t Ratio	Prob>|t|	Lower 95%	Upper 95%
Intercept	5.6452776	1.84302	3.06	0.0037*	1.9309151	9.3596401
Target Angle	0.9797401	0.018048	54.29	<.0001*	0.9433667	1.0161135

The results certainly seem to suggest that the sand scorpions are effective locators of their prey. The estimated slope is very close to 1.0, with a tight confidence interval. It is interesting that there appears to be an ever-so-slight, but statistically significant, bias indicated by the intercept of 5.6 degrees. (Could this be another case of handedness?)

What Have We Learned?

In this chapter we demonstrated the capabilities of JMP for performing standard inferential procedures with quantitative data. From stem (checking assumptions) to stern (making the inference), JMP guided the analysis seamlessly and naturally. Whether hypotheses were being tested or confidence intervals constructed the steps built naturally on what we had previously done with JMP when we considered basic exploratory data analysis techniques.

References

Andriessen, P., et al. (2008). Feasibility of noninvasive continuous finger arterial blood pressure measurements in very young children, aged 0–4 years. *Pediatric Research* 63(6):691–96.

Aydin, H., et al. (2004). Initial fixation strength of interference nail fixations for anterior cruciate ligament reconstruction with patellar graft (experimental study). *Knee Surgery, Sports Traumatology, Arthroscopy* 12(2): 94–97.

Brownell, P. H., and J. L. van Hemmen. (2001). Vibration sensitivity and a computational theory for prey-localizing behavior in sand scorpions. *American Zoologist* 41:1229–40.

Jones, D. S., and B. J. MacFadden. (1982). Induced magnetization in the monarch butterfly, *DANAUS PLEXIPPUS* (INSECTA, LEPIDOPTERA). *Journal of Experimental Biology* 96:1–9.

Roth, E. D. (2003). "Handedness" in snakes? Lateralization of coiling behavior in a cottonmouth, *Agkistrodon piscivorus leucostoma*, population. *Animal Behaviour* 66(2):337–41.

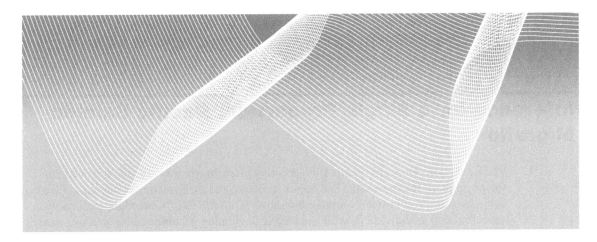

Chapter 7

Inference with Qualitative Data

Introduction

In this chapter, I will round out my discussion of inference, focusing on qualitative
(categorical) data. Once again, I will offer a sequence of vignettes for hypothesis testing
and confidence intervals. As we work through our examples of inference with categorical
data you will see that some of the techniques are slightly different from the standard

elementary textbook presentations in statistics. We will keep you aware of these differences as we proceed.

Inference for a Single Proportion: Pacific Salmon Migration

Pacific salmon (*Onocorhynchus spp.*) seemingly return from foraging in the ocean to their birthplace for spawning. The reason, process, and mechanisms of navigation are subjects of some controversy among biologists (Quinn, 1991). One might reasonably model salmon as individuals with a primitive navigation capability, able to sense where shallow water is but otherwise incapable of any navigation, and specifically unable to tell whether their home river is north or south of where they are when they reach shallow water. Such a salmon, swimming east, would find the coast and turn north or south randomly.

A competing theory is that salmon actually do have a navigation capability. Under this theory salmon would get close to shore and know whether to turn left or right to head for home. Studies in the laboratory have determined that the olfactory organs of fish are very sensitive to concentrations of chemicals in the water and that they remember the chemicals that identify their home rivers.

It is possible to test these competing claims by taking a "survey" of the salmon. When salmon are netted by fishing boats using gill nets, it is possible to determine the direction they are swimming by noting their orientation in the net when snagged. Jamon (1990) reasoned that salmon without navigational capabilities should be caught moving north or south in equal proportion. Alternatively, if salmon do have navigational capability, a fishing boat trolling north of the salmon's home river during migration should detect greater than half the salmon heading south. I will note in passing that this is not a simple salmon sampling problem. Could it be that salmon may be schooling? One might be suspicious of this with a large number of salmon gathered in one dip of the net. On the other hand, if few salmon were netted per dip, they could be regarded as independently sampled. In any case, we will trust the researcher and use Jamon's (1990) data to illustrate the use of JMP to perform inference for a single proportion. Let p = the population proportion of salmon traveling toward home, and test the hypothesis that $H_0 : p = 0.5$. In Jamon's (1990) sample 120 out of 200 netted salmon were detected swimming toward their home. The data are in the JMP file Salmon, and the data entry is shown in figure 7.1.

Figure 7.1 The data setup

	Direction	Count
1	Toward	120
2	Away	80

Entering data in a JMP table when working with categorical data is a great deal more efficient than using one row for each individual salmon; one row is used for each *category*. In the case of our salmon migration we have only two categories: the salmon that turned toward their home river or those that did not. Thus, we have two variables to enter: direction and the salmon count. (I note in passing that there are circumstances where individual salmon might be entered in a data table. The weight and length might also be measured as part of a larger study of migration habits.)

Notice that when you enter character data into the **Direction** column, JMP immediately knows that Direction is a categorical variable; the red bars next to the column name indicate this. To see the distribution of responses we use the Distribution Platform in JMP.

1. Select **Analyze → Distribution**.

2. Select **Direction → Y, Columns → Count → Freq**.

You should now see something similar to figure 7.2.

Figure 7.2 Distribution choices

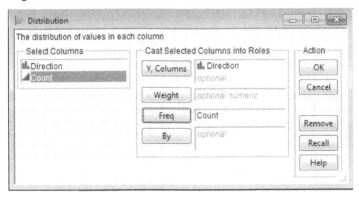

3. Click **OK**.

4. Click the **Distributions** hot spot and select **Stack**.

You should now see something similar to figure 7.3. Generally my preference is for a horizontal display, so I selected the **Distributions** hot spot and **Stack** to get figure 7.3. This is, of course, a judgment call. First, let's determine the 95 percent confidence interval for the proportion of homeward-bound salmon. We begin with the **Direction** context triangle:

Figure 7.3 Proportions, Stack display

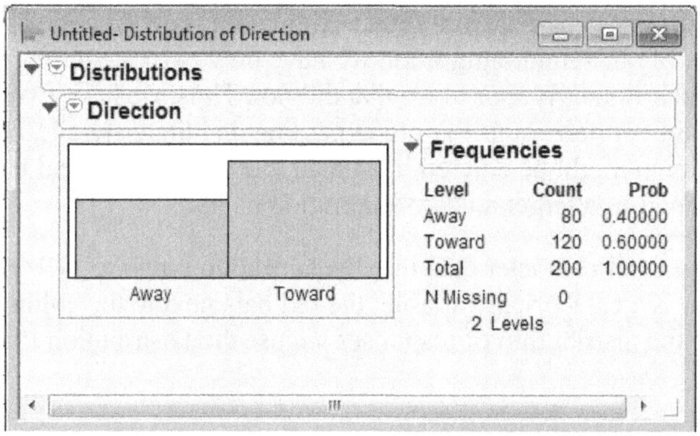

5. Select **Direction → Confidence interval → 0.95**.

JMP adds the confidence intervals to the panel as shown in figure 7.4. The confidence intervals here are slightly different from what your textbook formula and calculator may give you. As JMP notes, the confidence interval is what is known as a "score" confidence interval. It is well known that the large sample confidence interval (the "Wald" interval) for the population proportion is disappointing for some combinations of successes and sample sizes in that the confidence intervals depart significantly from the advertised probabilities of coverage of the true parameter (95 percent in our case). Elementary statistics books are beginning to recommend the "modified Wald," where one adds 2 to the numerator and denominator. The score interval is different from both the Wald and the modified Wald. The score interval works better for small samples but can be recommended for all sample sizes (Agresti and Coull, 1998). So that you may compare the results, the Wald confidence interval is (0.5321, 0.6679), and modified Wald is (0.5308, 0.6653).

Figure 7.4 Confidence intervals for the proportions

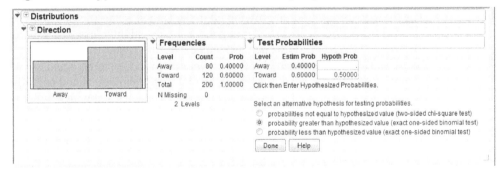

JMP does not know which of the values you have defined as a "success," so confidence intervals for each proportion, Away and Toward, are presented.

In this example, we have the natural null hypothesis, $H_0 : p = 0.5$. To test that hypothesis in JMP,

6. Select **Direction → Test Probabilities**.

This time JMP adds **Test Probabilities** to the panel, as we see in figure 7.5.

Figure 7.5 Hypothesis test selection for a proportion

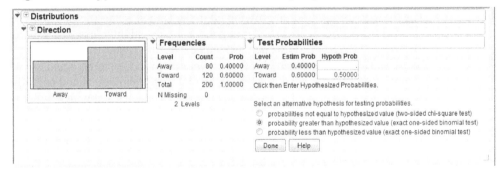

7. Enter 0.5 in the **Toward** row, select **probability greater than hypothesized value (exact one-sided binomial test) → Done**.

Notice again that JMP, using its computational power, is performing the "exact" test based on the binomial distribution, not the large sample approximation to the sampling distribution. The P-value on your calculator may differ slightly from that of JMP. My calculator shows a P-value of 0.0023, only slightly different from JMP's output, 0.0028, which is shown in figure 7.6.

Figure 7.6 Hypothesis test report for a proportion

Untitled- Distribution of Direction

▼ ▽ **Distributions**

▼ ▽ **Direction**

Frequencies			Test Probabilities		
Level	Count	Prob	Level	Estim Prob	Hypoth Prob
Away	80	0.40000	Away	0.40000	0.50000
Toward	120	0.60000	Toward	0.60000	0.50000
Total	200	1.00000			
N Missing	0				Hypoth
	2 Levels		Binomial Test	Level Tested	Prob (p1) p-Value
			Ha: Prob(p > p1)	Toward	0.50000 0.0028*

With a small P-value we can reject the hypothesis that salmon are randomly choosing their direction of travel when returning home to spawn. Of course, the statistics cannot tell us how the salmon are actually finding their way, but it appears that chance is not the mechanism.

Inference for Two Proportions: The Tooth Fairy

The lives of young children growing up in the United States include many figures, some real and some slightly less so. Among those figures slightly less so are monsters under the bed, and fantasy figures such as Santa Claus, the Easter Bunny, and the Tooth Fairy. Blair, McKee, and Jernigan (1980) were interested in the strength and duration of beliefs in fantasy figures as characteristics of the child's psychological and cognitive development. Specifically, they were interested in the ages at which belief in such fantasy figures declined.

The investigators interviewed white, middle-class, Christian children in southeastern Michigan and categorized them as either "firm believers" or "not firm believers" in various fantasy figures. The children's faith was of interest because Santa Claus and the Easter Bunny are associated with significant events in the Christian calendar. The data for belief in the Tooth Fairy is presented in table 7.1 and is in the file ToothFairy.

Table 7.1 Firm believers in the Tooth Fairy

Age (yrs)	Firm believers	Not firm believers
6-7	29	21
8-10	12	35

Our current interest centers on testing the hypothesis that the two population proportions are equal.

The data entry for inference with two proportions is very similar to the data entry for inference for a single proportion, as illustrated in figure 7.7. (I assigned Age to be a character variable in advance at data entry because JMP converts expressions such as "6–7" to dates using its two-digit year rule.)

Figure 7.7 Two-proportion data setup

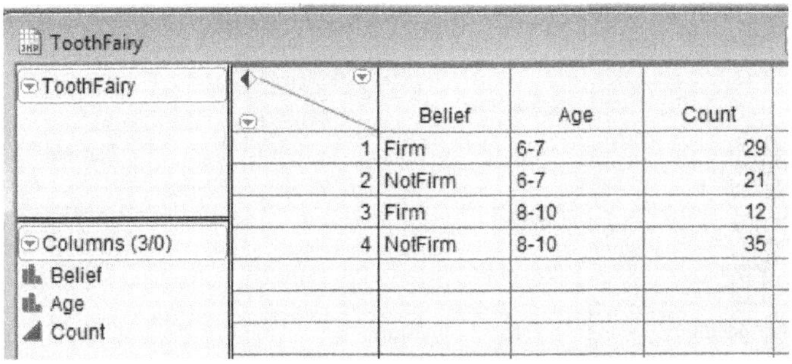

ToothFairy

ToothFairy

Columns (3/0)

Belief

Age

Count

	Belief	Age	Count
1	Firm	6-7	29
2	NotFirm	6-7	21
3	Firm	8-10	12
4	NotFirm	8-10	35

JMP does not provide the usual *z*-statistic for this hypothesis test; it calculates the more general Pearson chi-square statistic (more of which to follow). The Pearson chi-square statistic is the square of the *z* statistic in the case of two proportions and has the advantage of generalizing to more than two proportions.

1. Select **Analyze → Fit Y by X.**

We are wondering if age affects the prevalence of firm belief, so Belief is our response variable.

2. Select **Belief → Y, Response → Age → X, Factor → Count → Freq**.

You should see something similar to that shown in figure 7.8.

Figure 7.8 Two-proportion hypothesis test choices

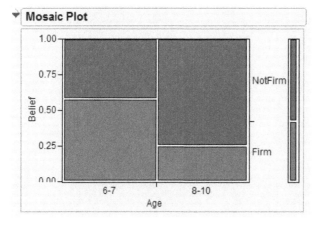

3. Click **OK**.

The plot in figure 7.9 is known as a mosaic plot. The bar on the right indicates the unconditional proportions for the two age groups; the two bars on the left indicate the conditional sample proportions for the 6–7-year-olds and the 8–10-year-olds. In this plot one can see very quickly that the proportions differ and get a visual sense of how much they differ.

Figure 7.9 Mosaic plot

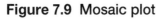

4. Click the **Contingency Table** hot spot.

5. Deselect **Total%**, **Col%**, and **Row%**.

6. Select **Expected**, **Deviation**, and **Cell Chi Square**.

A contingency table with information focused on the chi-square analysis is shown in figure 7.10. We can compare the expected and observed frequencies and see the contributions each cell makes to the chi-square statistic. As shown in figure 7.11, JMP calculates the Pearson chi square of 10.466 (which implies a z-statistic of 3.235). Notice that the P-values for all three alternative hypotheses are provided, together with a very clear verbal description of which P-value goes with which alternative hypothesis.

Figure 7.10 Chi-square analysis

Contingency Table

		Belief		
	Count Total % Col % Row %	Firm	NotFirm	
Age	6-7	29 29.90 70.73 58.00	21 21.65 37.50 42.00	50 51.55
	8-10	12 12.37 29.27 25.53	35 36.08 62.50 74.47	47 48.45
		41 42.27	56 57.73	97

Figure 7.11 Chi-square report

Tests

N	DF	-LogLike	RSquare (U)
97	1	5.3553402	0.0811

Test	ChiSquare	Prob>ChiSq
Likelihood Ratio	10.711	0.0011*
Pearson	10.466	0.0012*

Fisher's

Exact Test	Prob	Alternative Hypothesis
Left	0.9998	Prob(Belief=NotFirm) is greater for Age=6-7 than 8-10
Right	0.0011*	Prob(Belief=NotFirm) is greater for Age=8-10 than 6-7
2-Tail	0.0019*	Prob(Belief=NotFirm) is different across Age

The P-value of 0.0012 leads to the rejection of the hypothesis, and we conclude that the proportion of Firm Believers in the Tooth Fairy is less for Age 8–10 than for Age 6–7. Said another way, the proportion of Firm Believers appears to decrease between Age 6–7 and Age 8–10.

The Chi-Square Goodness of Fit: The Supreme Court

In recent years, appointments to the U.S. Supreme Court have been hot political potatoes. Vacancies on the court occur due to death, retirement, and—theoretically—impeachment, though this has never occurred. In a study of vacancies from 1837 to 1932 (a time over which the number on the court was nine), Wallis (1936) estimated the probabilities of the number of vacancies in any given year to be as indicated in table 7.2.

Table 7.2 Number of vacancies, 1837–1932

Number of vacancies in a year (1837-1932)	Probability
0	0.6065
1	0.3033
>1	0.0902

Cole (2010) gathered data on vacancies for the seventy-five years between 1933 and 2007 to see if any change had occurred in that period. He used as a baseline the probabilities calculated by Wallis (1936). These data are presented in table 7.3, the data entry is shown in figure 7.12, and the data are in the JMP file Supremes.

Table 7.3 Number of vacancies, 1933–2007

Number of vacancies in a year (1933-2007)	Observed
0	47
1	21
>1	7

Figure 7.12 Goodness of fit data setup

	NVacancies	Count
1	Zero	47
2	One	21
3	More	7

I will demonstrate the chi-square goodness of fit test using JMP and the data from Cole (2010). Data entry is similar to the salmon example:

1. Select **Analyze → Distribution → NVacancies → Y, Columns → Count → Freq → OK**.

2. Click the **Distributions** hot spot and select **Stack**.

3. Click the **NVacancies** hot spot and select **Test Probabilities**.

Our probabilities are theory-driven rather than estimated from the data, and we can enter them in the **Hypoth Prob** column shown in figure 7.13. We should note in passing that a common problem with decimal probabilities is round-off error. A quick check of the current probabilities shows that they add up to 1.000, but this is not always the case. Fortunately, JMP has a built-in capability to handle this problem.

Figure 7.13 Adjust probabilities for decimal round-off

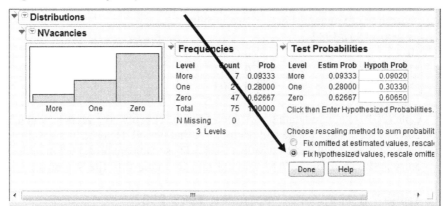

4.   Select the **Fix omitted at estimated values, rescale hypothesis** radio button (arrow). JMP will rescale the probabilities to be legal (that is, adding to 1.0).

5.   Enter the probabilities in the **Hypoth Prob** blanks and click **Done**.

Be careful; the order of variables is not the same as the order I had in the table! JMP alphabetizes the choices.

We see large P-values in figure 7.14, indicating that the data are consistent with the hypothesis of no change in the distribution of the number of vacancies per year in the modern (1933–2007) U.S. Supreme Court.

Figure 7.14 Goodness of fit report

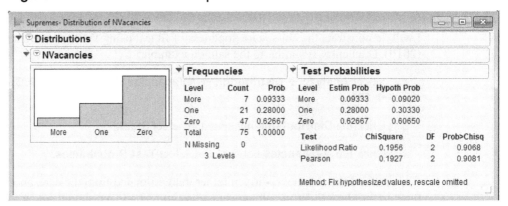

The Chi-Square Test of Independence: Race and Longevity in the Nineteenth Century

The nineteenth-century African-American experience is rich in anecdotal narrative and is typically reconstructed through the study of diaries, ledgers, and family records. Foster and Eckert (2003) used gravestones and burial records to "expand the historical understanding of an African American community in the rural Midwest in the nineteenth and twentieth centuries" through the examination of data (gravestones and written records) from Coles County, Illinois. They were able to determine the prevalence of African Americans, the ethnicity of surnames, ages at death, and to some extent the causes of death for almost 56,000 individuals, including 338 African Americans.

One part of their analysis focused on the mean age at death for blacks and whites. In most decades from the 1860s to the 1980s the age at death for whites was greater than that of blacks. In a breakdown of the data, the investigators coded the individuals as infants, children, adults, or elders for further analysis. Table 7.4 presents this breakdown.

Table 7.4 Race by age status at death

Age Status	Blacks	Whites
Infant	37	2140
Child	30	1230
Adult	123	4698
Elder	87	6417

Foster and Eckert hypothesized that African Americans who had achieved adult status historically did not live as long as whites, and that this explained most of the difference in mean ages across the ethnicities. Each individual's age status and ethnicity were determined as categorical variables. The data have been entered as shown in figure 7.15 and stored in the JMP file ColesCounty.

Figure 7.15 Data entry

	AgeStatus	Ethnicity	Count
1	Infant	Black	37
2	Child	Black	30
3	Adult	Black	123
4	Elder	Black	87
5	Infant	White	2140
6	Child	White	1230
7	Adult	White	4698
8	Elder	White	6417

ColesCounty

ColesCounty

Columns (3/0)
- AgeStatus
- Ethnicity
- Count

1. Select **Analyze → Fit Y by X → AgeStatus → Y, Response → Ethnicity → X, Factor → Count → Freq → OK**.

2. Click the **Contingency Table** hot spot.

3. Hold down the **Alt** key and deselect **Total%**, **Col%**, and **Row%**; select **Expected**, **Deviation**, and **Cell Chi Square**.

Because the proportion of African-American graves is so small, it may be necessary to enlarge the mosaic plot to see those proportions on the left of the plot (see figure 7.16).

Figure 7.16 Mosaic plot

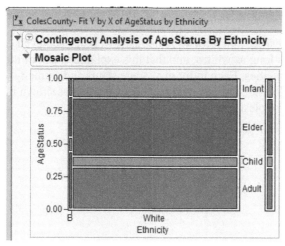

The chi-square analysis is shown in figures 7.17 and 7.18. The very small P-value is surely due to the huge sample size, and we will probably wish to consider the actual proportions in a thoughtful analysis of the data.

Figure 7.17 Contingency table

Contingency Table

		AgeStatus				
Ethnicity	Count Total % Col % Row %	Adult	Child	Elder	Infant	
	Black	123 0.83 2.55 44.40	30 0.20 2.38 10.83	87 0.59 1.34 31.41	37 0.25 1.70 13.36	277 1.88
	White	4698 31.82 97.45 32.43	1230 8.33 97.62 8.49	6417 43.47 98.66 44.30	2140 14.50 98.30 14.77	14485 98.12
		4821 32.66	1260 8.54	6504 44.06	2177 14.75	14762

Figure 7.18 Chi-square report

Tests

N	DF	-LogLike	RSquare (U)
14762	3	12.052903	0.0007

Test	ChiSquare	Prob>ChiSq
Likelihood Ratio	24.106	<.0001*
Pearson	24.293	<.0001*

The proportions can be seen in the contingency table in figure 7.19; they are shown as the row percentages, the last value in each cell. The analysis by age status sheds some light on Foster and Eckert's hypothesis of a shorter life for African Americans. They observe that the proportions of children and infant deaths were very similar for blacks and whites, suggesting that the deaths were driven by disease and the hardships of frontier life. There is a noticeable difference for those who achieved adulthood; the proportion of elderly blacks is significantly smaller than the proportion of elderly whites, consistent with Foster and Eckert's theory.

Figure 7.19 Table proportions

⊽ Contingency Table

		Adult	Child	Elder	Infant	
	Count	Adult	Child	Elder	Infant	
	Expected					
	Deviation					
	Cell Chi^2					
	Black	123	30	87	37	277
		90.4631	23.6431	122.044	40.8501	
		32.5369	6.35686	-35.044	-3.8501	
		11.7025	1.7092	10.0624	0.3629	
Ethnicity	White	4698	1230	6417	2140	14485
		4730.54	1236.36	6381.96	2136.15	
		-32.537	-6.3569	35.0436	3.85009	
		0.2238	0.0327	0.1924	0.0069	
		4821	1260	6504	2177	14762

The Chi-Square Test of Homogeneity of Proportions: Is Golf for the Birds?

Because of the popularity of golf, new courses are opening all over the world. LeClerc and his colleagues (2005) sought to assess the ecological impact of these courses. Potential ecological problems include habitat fragmentation, chemical pollution due to pesticides, and loss of native vegetation. The focus of their study was eastern bluebirds (*Sialia sialis*), a species often abundant on golf courses. These bluebirds are "secondary cavity nesters;" that is, they do not excavate their own cavities for nesting. They are particularly attracted to birdhouses and other nesting structures put up in backyards. LeClerc and colleagues monitored the lives of these birds during a breeding season (1 April–30 August) at nine golf courses and ten non-golf course sites. The non-golf course sites had habitat similar to the golf courses, but with no known pesticide use. One characteristic of interest was reproductive success, as measured by nest density in nest boxes. If nest boxes on golf courses are less attractive to bluebirds, this should show up as a difference in the distribution of occupancy frequencies. The data on nest boxes occupied only by bluebirds are presented in table 7.5.

Table 7.5 Nest boxes by location

Site	0 nests	1 nest	2 -3 nests
Golf	48	80	55
Non-golf	62	74	26

Our data have been entered as shown in figure 7.20 and stored in the file Golf.

Figure 7.20 The data setup

	Site	NNests	Count
1	Golf	Zero	48
2	Off	Zero	62
3	Golf	One	80
4	Off	One	74
5	Golf	Two-Three	55
6	Off	Two-Three	26

1. Select **Analyze → Fit Y by X → NNests → Y, Response → Site → X, Factor → Count → Freq → OK**.

2. Hold down the **Alt** key and click the **Contingency Table** hot spot.

3. Deselect **Total%**, **Col%**, and **Row%**; select **Expected**, **Deviation**, and **Cell Chi Square**.

JMP presents the information, as shown in figures 7.21, 7.22, and 7.23.

Figure 7.21 Mosaic plot

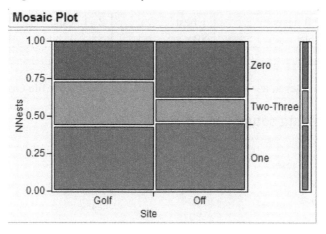

Figure 7.22 Contingency table

Contingency Table

			NNests		
	Count Total % Col % Row %	One	Two-Three	Zero	
Site	Golf	80 23.19 51.95 43.72	55 15.94 67.90 30.05	48 13.91 43.64 26.23	183 53.04
	Off	74 21.45 48.05 45.68	26 7.54 32.10 16.05	62 17.97 56.36 38.27	162 46.96
		154 44.64	81 23.48	110 31.88	345

Figure 7.23 Chi-square analysis

Tests

N	DF	-LogLike	RSquare (U)
345	2	5.6790923	0.0155

Test	ChiSquare	Prob>ChiSq
Likelihood Ratio	11.358	0.0034*
Pearson	11.161	0.0038*

The mosaic plot presents the story graphically. (Notice that the order for NNests is perhaps not what one might like. The order of the values as presented can be altered using the **Value Ordering** property; consult JMP Help for details.) The proportions of single-nest nest boxes are very similar for golf courses and non-golf course sites. The proportion of nest boxes with zero nests is less, and the proportion of nest boxes with two or three nests is greater than the golf courses' proportion. The P-value certainly suggests a statistically significant result! It appears that the golf sites were more attractive nesting grounds than non-golf course sites.

For those considering building a golf course in the near future, we note that on many of the other variables measured in this study, the golf courses seemed to be more bluebird-friendly than the non-golf course sites.

What Have We Learned?

In this chapter we demonstrated the facility with categorical data in JMP. Both proportions (hypothesis testing and confidence intervals) and the chi-square procedures (goodness of fit, independence, and homogeneity of proportions) were considered.

References

Agresti, A., and A. Coull. (1998). Approximate is better than "exact" for interval estimation of binomial proportions. *American Statistician* 52(2):119–26.

Blair, J. R., J. S. McKee, and L. F. Jernigan. (1980). Children's belief in Santa Claus, Easter Bunny and Tooth Fairy. *Psychological Reports* 46:691–94.

Cole, J. H. (2010). Updating a classic: "The Poisson distribution and the supreme court" revisited. *Teaching Statistics* 32(3):78–80.

Foster, G., and C. Eckert. (2003). Up from the grave: A sociohistorical reconstruction of an African American community from cemetery data in the rural Midwest. *Journal of Black Studies* 33(4):468–89.

Jamon, M. (1990). A reassessment of the random hypothesis in the ocean migrations of Pacific salmon. *Journal of Theoretical Biology* 143:197–213.

LeClerc, J. E., et al. (2005). Reproductive success and developmental stability of eastern bluebirds on golf courses: Evidence that golf courses can be productive. *Wildlife Society Bulletin* 33(2):483–93.

Quinn, T. P. (1991). Models of Pacific salmon orientation and navigation on the open ocean. *Journal of Theoretical Biology* 150:539–45.

Wallis, W. A. (1936). The Poisson distribution and the Supreme Court. *Journal of the American Statistical Association* 31(194):376–80.

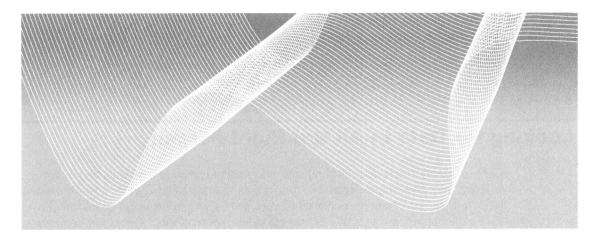

Chapter 8

Importing Data into JMP

Introduction

Focusing your attention on the obvious, I will make two assertions: (1) statistics is about data, and (2) the availability of data in our connected world is nothing less than astounding. "Data" are now available in many forms: images, shape files, graphs, and tables in scientific publications, and of course in the usual row and column format. Raw data are available in large quantities in spreadsheets, databases, and other traditionally statistical formats and can be easily accessed with JMP. Many Web pages present data in a text or table format that can be copied and pasted into JMP with little or no prior editing. (More on this in chapter 9.)

In this chapter I discuss and demonstrate how to import data into JMP.

Looking for Data in All the Right Places

Deep in the processor(s) of the computer, information is read in a manner not all that different from how humans read. We form words from alphabetic symbols; the computer takes "bits" of information, conceived of as 0s and 1s, and forms larger chunks of information known as bytes. We extract meaning from words and syntax; the computer reacts to sequences of bytes, some in the form of instructions, some in the form of data. Our phrases, sentences, and paragraphs are structured by a language to provide meaning. The computer's programs and data must also be planned in advance to facilitate the construction of meaning from bits and bytes. Humans have flexibility in the arrangement and spelling of words and in the format of presentation, and in the case of spoken communication variations of accent are allowed without loss of meaning. Here the similarity between computer and human ends; computers have only the flexibility that human ingenuity can program into them. Human ingenuity has so far succeeded in programming a seemingly remarkable flexibility in the computer's capability to "read" data, but it is built on a thin veneer of shallow capability, executed by very fast computers.

When data is entered into JMP and saved electronically, JMP stores the data in a very precisely defined format for future reading. Other computer programs, such as Microsoft Word or Excel, not having been programmed to read JMP files, may be able to read the bytes, but they are unable to properly interpret them. To get a sense of how Word, for example, interprets JMP files, ask Word to read one. I did, with results shown in figure 8.1. Other than a small amount of some accidentally readable characters in the file, Word could not quite figure out how to read the JMP file. JMP has to solve the same problem when attempting to read, or "import," data from files that are not JMP. Over the years of software development different programs, using data but built for essentially nonstatistical purposes, have defined formats consistent with their purpose. The JMP software folks have designed into JMP the capability of reading many of these types of files stored in other formats. In addition, "industry standard" data file formats have grown to facilitate the transfer of data among different data producers and consumers. When JMP reads a file that is not a JMP file, we say that JMP is "importing" the file. Which files you will be able to import into JMP will depend to some extent on the software you have installed on your machine.

Figure 8.1 Word reading a JMP file

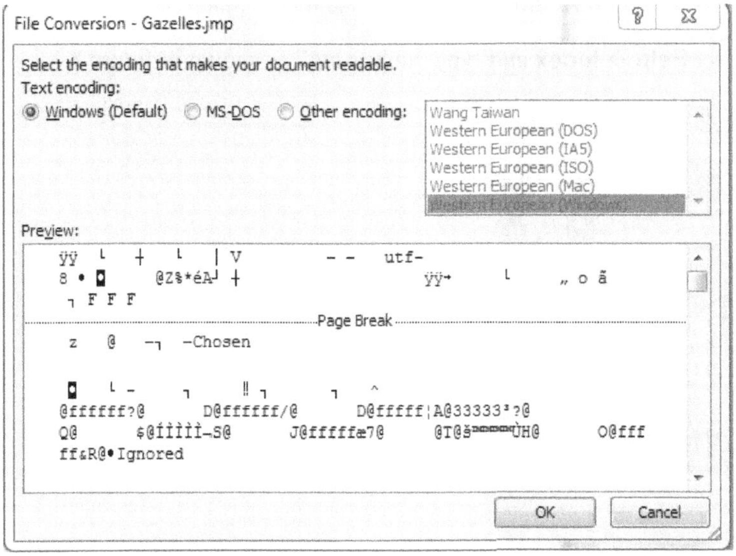

1. Select **File → Open**.

You should see a drop-down menu indicating the types of files to choose from. (Note: Mac users will see the available file types highlighted by their file name. On the Mac, JMP can open any highlighted file.) In my case the present choice is **All JMP Files**. Click on this drop-down menu to see your import options. Mine are shown in figure 8.2.

Figure 8.2 Readable file formats

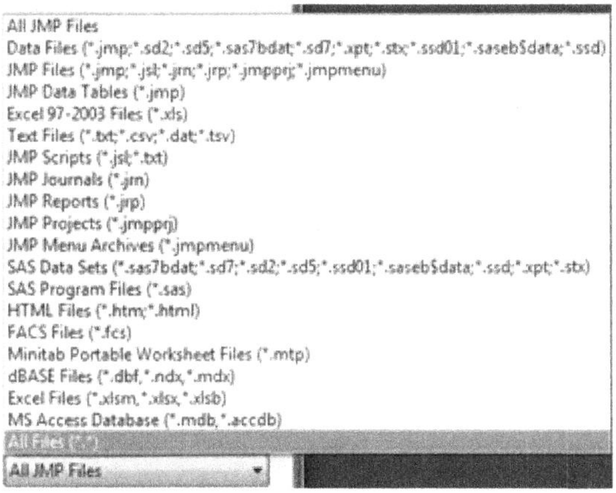

The Help capability in JMP gives excellent detailed instructions about importing files of different types, and I encourage you to browse these capabilities.

2. Select **Help → Index** and type "importing" as shown in figure 8.3.

Figure 8.3 Help choices

In an elementary statistics class, most of the files that will be imported will probably be Excel files and text files, since these types of files are almost universally available on computers. To illustrate downloading and reading files with JMP, I will give you the actual URLs and walk you through the steps for importing the data. I will not analyze these data statistically, but of course you are very welcome to do so by practicing your new JMP skills.

Importing Microsoft Excel Spreadsheets

To illustrate the import of Excel files, I use a file downloaded from the U.S. Department of Energy website, http://www.fueleconomy.gov/feg/download.shtml. The fuel economy data from the 2011 Datafile contain information about vehicle testing done at the Environmental Protection Agency's National Vehicle and Fuel Emissions Laboratory in Ann Arbor, Michigan, and by vehicle manufacturers with oversight by the EPA.

Importing data from an Excel file proceeds in a manner very similar to opening JMP files.

1. Click **File → Open**, and in the drop-down menu, select **Excel 97-2003 Files (*.xls)**.

2. Select **2011FEguide-public.xls**.

An issue that must be considered when importing Excel files is whether the first row consists of column names. Notice the **Always enforce Excel row 1 as labels** options in figure 8.4. Some Excel files (as well as text files) will include a first row with variable names, while some will not. If JMP believes the first row has variable names, it will import the first row as column headings, not data. If in fact there are no variable names in the first row, JMP will treat the first row of data as column headings. If the Excel file *does* have variable names in the first row, but JMP believes otherwise, the variable names will be imported as the first row of data, also something not desirable. The best strategy is probably to view the file in Excel, take note of what is in the first row, and then select **Always** or **Never**.

Figure 8.4 Choosing the best guess

The EPA data has variable names in the first row.

3. Select the **Always** button and **Open**.

JMP will respond by reading the Excel file, part of which is shown in figure 8.5.

Figure 8.5 Imported Excel file 20011FEguide-public

		Model Yr	Mfr Name	Division	Carline	Verify Mfr Cd	Index (Model Type Index)	Eng Displ	# Cyl
	1	2011	Audi	Audi	TT ROADSTER QUATTRO	ADX	26	2	4
	2	2011	Audi	Audi	R8 Spyder	ADX	27	5.2	10
	3	2011	Audi	Audi	R8	ADX	28	5.2	10
	4	2011	Audi	Audi	R8 Spyder	ADX	29	5.2	10
	5	2011	Audi	Audi	R8	ADX	30	5.2	10

Copying and Pasting Microsoft Excel Spreadsheets

Excel spreadsheets and portions of Excel spreadsheets can also be transferred via copying and pasting. To utilize copying and pasting, you must have both JMP and Excel open.

1. In Excel, select the rows and columns that you wish to copy and select **Edit →
   Copy**. (Or, in Windows 7, just **Copy**.)

2. In JMP, create a new data table by selecting **File → New**.

3. In JMP, hold down the **Shift** key and select **Edit → Paste** or **Edit → Paste with
   Column Names**, depending on the status of the first row.

Importing Text Files

Another common format used for transfer of data is a text format. Text formats can be
identified by their file extensions: .txt, .csv, .tsv, or .dat. The .csv and .tsv extensions
primarily indicate comma-separated variables and tab-separated variables format; .txt and
.dat generally indicate human-readable data as opposed to data that can only be read by
specific computer software. The .csv and .tsv formats are available with most database
programs and Excel. The .csv format consists of records of data, one line per record, with
variables delimited by commas. In some cases commas may be part of the data, in which
case the data would be enclosed in quotation marks. Tabs are less common in data, and
for this reason .tsv is perhaps preferable to the .csv format. (If tabs do appear in data for
some unanticipated reason, the data value is also surrounded by quotation marks.) The
.txt and .dat formats are more generic, but if the contents are intended to be data for
statistical analysis it is quite likely that a row and column format similar to .csv or .tsv is
being used.

Importing data from a .txt or .dat text file can be a bit more complicated than importing
an Excel file because some combination of you and JMP need to discern some
characteristics of the file format. The .csv and .tsv formats are usually dependably row
and column, but different computer systems (PC, Mac, Linux, or even a mainframe) will
differ in how they indicate an end-of-line.

1. Select **File → Open**, and in the drop-down menu, select **Text Files
   (.txt;.csv;*.dat;*.tsv)**.

There are some options to consider, as shown in figure 8.6.

Figure 8.6 Choosing text import preferences

These options represent different hints to JMP about the file format. Easiest to use is the **Data, using best guess** option. This is not the easiest because you or I guess the format; it is easiest because JMP guesses the format. JMP is excellent at this discernment of file formats and should be regarded as the fellow expert in the matter. The next-best option is for you or I to guess, using the **Data with Preview** option. Even then, JMP will be at our side, providing expert guidance. The **Data, using Text Import preferences** option allows you to specify more options for JMP and would generally be used if you had a more complex data structure than rows and columns, possibly with more than a single row of column headers.

To accentuate the different approaches we will step through each of these options using a single data set, downloaded from earthquake.usgs.gov, a website of the U.S. Geological Survey devoted to earthquakes and other natural hazards. The data we will use are locations and measures of earthquakes having at least a magnitude of 2.5 in United States and adjacent areas, and earthquakes anywhere in the world with at least a magnitude of 4.5. To better appreciate what JMP has to work with when making its guesses, you might open the QuakeData file with your word processor. QuakeData is a text file, so you may have to set your word processor to display "All Files *.*". If you would like to try downloading data and starting at that step, point your browser to http://earthquake.usgs.gov/earthquakes/catalogs/ and under CSV Files pick the "5+" earthquakes for the past seven days. Be aware, however, that the last seven days on your calendar and my last seven days as I write this will differ!

Here are the first few lines of the file for my last seven days:

```
Src,Eqid,Version,Datetime,Lat,Lon,Magnitude,Depth,NST,Region
us,c0003vmn,7,"Tuesday, May 31, 2011 16:26:12
UTC",40.2975,143.1563,5.0,28.80,157,"off the east coast of
Honshu, Japan"
us,c0003vjh,7,"Tuesday, May 31, 2011 12:28:36
UTC",39.3997,141.9330,5.6,40.00,295,"eastern Honshu, Japan"
us,c0003uav,8,"Monday, May 30, 2011 00:06:27 UTC",-15.1482,-
173.4925,5.2,27.20,190,"Tonga"
us,c0003tx6,8,"Sunday, May 29, 2011 18:24:01 UTC",-
7.7506,101.7467,5.9,9.50,58,"southwest of Sumatra, Indonesia"
```

```
us,c0003ttb,8,"Sunday, May 29, 2011 13:44:20 UTC",-
6.5907,129.7977,5.2,146.90,78,"Banda Sea"
us,c0003tpw,8,"Sunday, May 29, 2011 07:33:13 UTC",-
5.8344,149.3044,5.4,113.00,61,"New Britain region, Papua New
Guinea"
```

As you can see, the original file contains a variety of different sorts of data: numeric, alphabetic, and mixed, with interesting interspersed punctuation for human readability. Note that in this set of data the first row contains variable names. Choose **Data, using best guess**, navigate to your data folder, and open the file QuakeData.

Figure 8.7 The best guess from JMP

Src	Eqid	Version	Datetime	Lat	Lon	Magnitude	Depth	NST	Region
us	c0003vmn	7	Tuesday, May 31, 2011 16:26:12 UTC	40.2975	143.1563	5	28.8	157	off the east coast of Honshu, Japan
us	c0003vjh	7	Tuesday, May 31, 2011 12:28:36 UTC	39.3997	141.933	5.6	40	295	eastern Honshu, Japan
us	c0003uav	8	Monday, May 30, 2011 00:06:27 UTC	-15.1482	-173.4925	5.2	27.2	190	Tonga
us	c0003tx6	8	Sunday, May 29, 2011 18:24:01 UTC	-7.7506	101.7467	5.9	9.5	58	southwest of Sumatra, Indonesia
us	c0003ttb	8	Sunday, May 29, 2011 13:44:20 UTC	-6.5907	129.7977	5.2	146.9	78	Banda Sea
us	c0003tpw	8	Sunday, May 29, 2011 07:33:13 UTC	-5.8344	149.3044	5.4	113	61	New Britain region, Papua New Guinea

The first few rows as JMP has guessed them appear in figure 8.7. Not surprisingly, JMP has discerned the nature of the variables and placed them correctly in rows and columns. Notice in figure 8.7 that JMP has detected some character values in the variables, "Eqid" and "version." When downloading files this information can be an early warning sign of data or transmission errors. As an example, if one expected "version" to be a numeric variable, seeing that JMP has categorized it as a character variable may indicate some sort of problem, possibly with the data entry or with the data transfer. In these data, an "A" in the **Version** column indicates a particular mathematical earthquake model used to estimate the magnitude (Olsen, personal communication). When data are imported in JMP, numeric data is treated as continuous by default, and any column with non-numeric characters is deemed to be categorical. JMP will interpret the numeric values of Version as categorical. Sometimes a character value such as a "*" may be used to indicate "missing" data in a data file. JMP would flag this as indicating categorical data. Your response in that case would be twofold. First, simply delete the "*" in the table—JMP would interpret that missing value as, well, a missing value. After deleting any of these that you find, you would then have to inform JMP that the column contains numeric, not categorical, data. (I will show you how to do this later in this chapter when we consider data on the salaries of the New York Yankees.)

Now that we have an idea about the capabilities of JMP, let us consider our own capabilities. Close the QuakeData file and open it again, this time with the **Data with Preview** option selected.

1. Select **File → Open**, and in the menu, select **Text Files (.txt;.csv;*.dat;*.tsv)**.

2. Select **Data with Preview → Open**.

You should see the display in figure 8.8. Remember, if you are following along after downloading the current seven-day file your data will differ from mine.

Figure 8.8 JMP guesses with data with preview

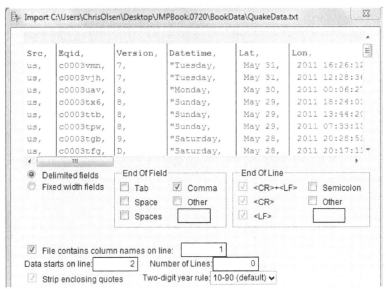

An important thing to notice here is that the best guess from JMP about the appropriate data table is displayed for your consideration. Use the scroll bars, both horizontal and vertical, to check out the data. If you don't see any problems, click **Next** and then **Import**, and you are on your way to analyzing the data. You need intervene only if you see something weird in the data table. If JMP sees a problem of some kind it will try to alert you. As an example, use your word processor and open the file QuakeDataWSrc. This file contains the earthquake data, plus the URLs for my reference. The first few lines look like this:

```
http://earthquake.usgs.gov/earthquakes/catalogs/
http://earthquake.usgs.gov/earthquakes/catalogs/eqs7day-M1.txt
Src,Eqid,Version,Datetime,Lat,Lon,Magnitude,Depth,NST,Region
us,c0003vmn,7,"Tuesday, May 31, 2011 16:26:12
UTC",40.2975,143.1563,5.0,28.80,157,"off the east coast of
Honshu, Japan"
us,c0003vjh,7,"Tuesday, May 31, 2011 12:28:36
UTC",39.3997,141.9330,5.6,40.00,295,"eastern Honshu, Japan"
us,c0003uav,8,"Monday, May 30, 2011 00:06:27 UTC",-15.1482,-
173.4925,5.2,27.20,190,"Tonga"
```

The preview in JMP for this file is shown in figure 8.9. Inspection of this preview gives you a clear indication of the problem, in this case a few lines of clutter before the actual data. When faced with this problem the easiest thing to do is to use your word processor and delete the offending early lines.

Figure 8.9 "Corrupt" data: The preview

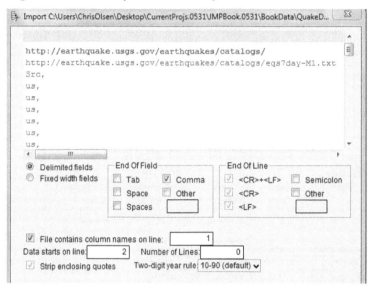

Dealing with a corrupt file is a bit more problematic and is something of an art, but JMP will, as always, assist you in the process. The number of ways that a data file can go wrong begins with data entry and continues with any number of possible human or electronic anomalies. The key concept when "cleaning the data" is that you know what it is supposed to look like in the data table. If you aren't seeing what you know should be there, you need to fix it! As an example, I will take the second data line of the earthquake data and delete the comma between "295" and "eastern Honshu, Japan." The file with the data changed as indicated is a text file named CorruptQuakeData. When JMP opens the file and you scroll through the data, you will immediately see the anomaly shown in figure 8.10.

Figure 8.10 More "corrupt" data: The preview

Once problems such as these are detected, it is usually a simple matter to fix them with your word processor.

Sometimes files created by individuals for specific purposes, or data gathered from physical measuring devices for direct computer processing, will contain the moral equivalent of rows and columns of data, but are formatted in some nonstandard manner. Or possibly you may latch onto a data file created by a different computer system that uses characters other than commas to separate data values and/or characters other than a carriage-return-linefeed combination. This is less of a problem with more recently archived data files, but it can occur. If you frequently acquire data in one of these different formats you may wish to use the **Data, using Text Import preferences** option, shown in figure 8.11. The bookended steps to make your choices are easy.

Figure 8.11 Choosing your own format

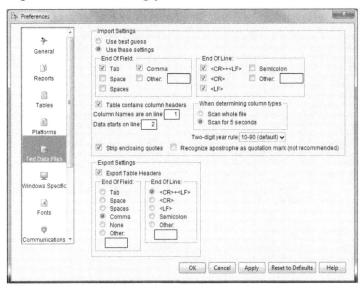

1. Select **File → Preferences → Text Data Files**.

2. Make appropriate choices from the panel shown in table 8.1.

3. Click **OK**.

Table 8.1 Options for file formatting

Box	Explanation
End of Field	What character indicates the end of the data value?
End of line	What character indicates the end of a record, or case?
Table contains column headers	What line are the column names on? On what line does the data begin?
When determining column types	How far should JMP look for data other than numeric-looking (i.e. non-numeric symbols)?
Two-digit year rule	How are dates formatted in the data file?
Strip enclosing quotes	Should quote marks be removed in the data? (Only used with fixed-width files, to be considered shortly.)
Recognize apostrophe as quotation mark	Does the data source use an apostrophe for quotation marks?
Export settings	What format parameters do you want to use when you send files to others?

Those steps in-between are the ones where the important decisions are made, and these decisions depend on the format of the file you are attempting to import. Generally, you would already know the correct answers because you created the file yourself or you acquired format information from the same source as the data.

Importing Data from the Internet: HTML Tables

Finally, let's consider the capability within JMP of downloading data contained in HTML table files on the Internet. Web pages can contain data in the form of rows and columns of cells known as an HTML table. In HTML one creates these tables using the table tag, <table>. Subsequent tags define the structure of the table. HTML tables can contain all sorts of things: text, images, and data. Our interest is, of course, data.

For JMP purposes we are interested in data that is structured in a table of rows and columns that can be interpreted as observations of variables, or at least contain a subset with that structure. HTML tables can be structured with different numbers of columns in different rows, and—bottom-line—not every HTML table represents a data set as we understand them. The message here is that before you attempt to import data from an HTML table on a web page, check to be sure it contains a table of data, as opposed to just a table.

As it happens, I have a website in mind that contains lots of tables, not all of which are data tables.

1. Select **File → Internet Open**.

The web page, http://en.wikipedia.org/wiki/World_population, contains information on the population of the world and many tables of data. Enter the URL into the **Internet Open** dialog box as shown in figure 8.12.

Figure 8.12 Internet open

2. Select **OK**.

The **Extract HTML Table to JMP File** option searches the web page for HTML tables and presents a list, as shown in figure 8.13. Notice that the ever-helpful JMP is (a) warning you about some tables not containing data as JMP understands the term, and (b) pointing out the most likely candidates for JMP-compliant HTML tables.

Figure 8.13 Data tables

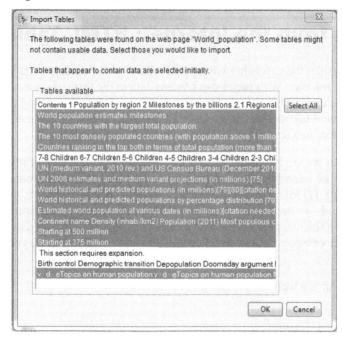

I selected the "10 countries with the largest total population" and was rewarded with the JMP file shown in figure 8.14. Well, it isn't actually a JMP file until it is saved, but you get the idea.

Figure 8.14 The 10 countries with the largest populations

	Rank	Country / Territory	Population	Date	% of world population	Source
1	1	People's Republic of China[57]	1345250000	July 21, 2011	0.194	Chinese Official Population Clock
2	2	India	1210193422	March 2011	0.17	Census of India Organisation
3	3	United States	311799000	July 21, 2011	0.045	United States Official Population Clock
4	4	Indonesia	238400000	May 2010	0.0338	SuluhNusantara Indonesia Census report
5	5	Brazil	194973000	February 21, 2011	0.0281	Brazilian Official Population Clock
6	6	Pakistan	176690000	July 21, 2011	0.0255	Official Pakistani Population Clock
7	7	Bangladesh	164425000	2010	0.0237	2008 UN estimate for year 2010
8	8	Nigeria	158259000	2010	0.0228	2008 UN estimate for year 2010
9	9	Russia	141927297	January 1, 2010	0.0205	Federal State Statistics Service of Russia
10	10	Japan	127380000	June 1, 2010	0.0184	Official Japan Statistics Bureau

Sometimes HTML tables are only almost usable. As an example, consider the salaries of major league baseball teams at http://espn.go.com/mlb/team/salaries/_/name/nyy/new-york-yankees (see figure 8.15).

Figure 8.15 Partial web page

Roster	Lineup	Salaries

PLAYER SALARIES				TEAM SALARIES		
RK	PLAYER	Salary (US$)		RK	PLAYER	Salary (US$)
1	CC Sabathia	24,285,714		1	New York Yankees	196,854,630
2	Mark Teixeira	23,125,000		2	Philadelphia Phillies	172,976,381
3	A.J. Burnett	16,500,000		3	Boston Red Sox	160,257,476
4	Mariano Rivera	14,911,701		4	Los Angeles Angels	138,999,024
5	Derek Jeter	14,729,365		5	Chicago White Sox	129,285,539
6	Jorge Posada	13,100,000		6	Chicago Cubs	126,380,663
7	Robinson Cano	10,000,000		7	New York Mets	120,147,311
8	Nick Swisher	9,100,000		8	San Francisco Giants	117,784,333

Partial results of choosing the **Extract HTML Table to JMP File** option are shown in figure 8.16. JMP pulled in the first two rows, the titles Rank and Player, naturally enough, since they were a part of the HTML table. This identifying information is of no interest to us (except for providing variable names).

Figure 8.16 The team salaries imported into JMP

	Column 1	Column 2	Column 3
1	Team Salaries	Team Salaries	Team Salaries
2	RK	PLAYER	Salary (US$)
3	1	New York Yankees	196,854,630
4	2	Philadelphia Phillies	172,976,381
5	3	Boston Red Sox	160,257,476
6	4	Los Angeles Angels	138,999,024
7	5	Chicago White Sox	129,285,539
8	6	Chicago Cubs	126,380,663
9	7	New York Mets	120,147,311
10	8	San Francisco Giants	117,784,333
11	9	Minnesota Twins	112,737,000

Click and drag on the first two rows, and then make the following selection.

1. Select **Rows → Delete Rows**.

There is another slight problem: JMP interpreted the contents of the first two rows as containing categorical rather than numeric data. (Recall that you can tell this by looking at the bar chart icon at the left of the panel.) The data of interest are in the salaries in Column 3. We need to inform JMP that these values are to be interpreted as numbers (see figure 8.17).

Figure 8.17 New York Yankee salaries

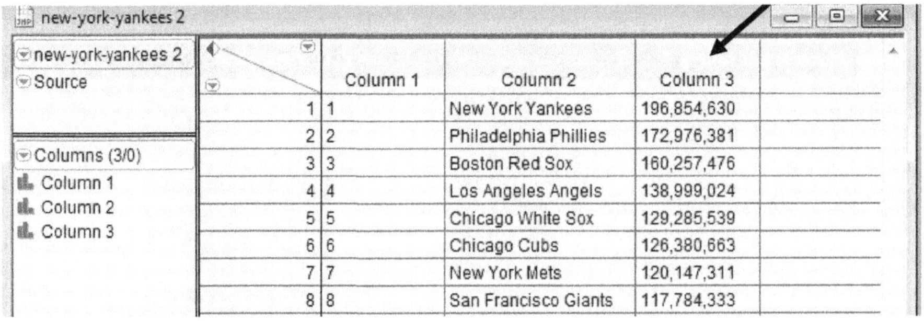
new-york-yankees 2
new-york-yankees 2
Source
Columns (3/0)
Column 1
Column 2
Column 3
Column 1
Column 2
Column 3
1 1 New York Yankees 196,854,630
2 2 Philadelphia Phillies 172,976,381
3 3 Boston Red Sox 160,257,476
4 4 Los Angeles Angels 138,999,024
5 5 Chicago White Sox 129,285,539
6 6 Chicago Cubs 126,380,663
7 7 New York Mets 120,147,311
8 8 San Francisco Giants 117,784,333

Double-click on "Column 3" to bring up the **Column** panel shown in figure 8.18. Make the changes indicated there and click **OK.** (If the only change needed for a variable is the modeling type, you can click on the icon to the left of "Column 3" in the **Columns (3/0)** panel and make the change there.)

Figure 8.18 Yankee Salaries—Column Panel

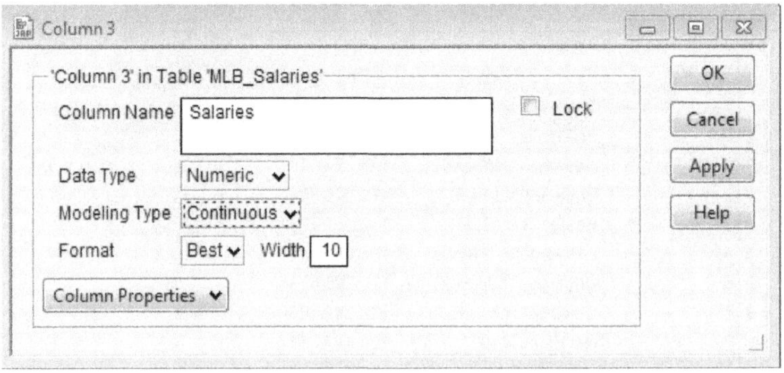
Column 3
'Column 3' in Table 'MLB_Salaries'
Column Name
Salaries
Lock
Data Type
Numeric
Modeling Type
Continuous
Format
Best
Width
10
Column Properties
OK
Cancel
Apply
Help

With these changes, we have successfully converted a problematic HTML table into working JMP format. After saving the data in a JMP file for safekeeping, we are ready to analyze.

What Have We Learned?

In this chapter we learned a bit about non-JMP file formats and how to convert them into a format for analysis with JMP. JMP works hand-in-glove with non-JMP files, although some human intervention is needed, on occasion, to facilitate the conversion to the correct format in JMP.

Reference

Olsen, A. (2011). Personal communication by telephone, August 30, 2011.

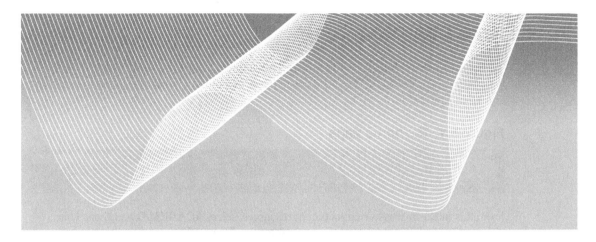

Chapter **9**

JMP on the Web

Introduction

In this chapter, I introduce you to a couple of examples that represent the continuing commitment by JMP to its users: the Interactive Learning Tools and the Learning Library. These are resources primarily designed for teachers and learners of statistics and are a small subset of resources available to the vast and varied JMP community.

Interactive Learning Tools

To begin to use the Interactive Learning Tools, direct your browser to the JMP website, www.jmp.com. You should see the home page, and at the top please notice the section shown in figure 9.1.

Figure 9.1 The home page

From this rather impressive collection of resources, select **ACADEMIC** and focus your attention on the list at the left of the page, shown in figure 9.2.

Figure 9.2 Our focus

Academic Licensing

International

See JMP

Instructors

Students

Researchers

Interactive Learning Tools

Learning Library

Books and JMP

Now select **Interactive Learning Tools**. You will be greeted with what is shown in figure 9.3, which is a partial list of some demonstration modules written for the classroom. These modules are JSL scripts, that is, computer programs written to be executed by JMP. You can execute them by double-clicking on the files or using the **File → Open** sequence from the JMP menu. These JMP scripts have been written for teachers and students and can be used by teachers in a demonstration mode or by students as a computer lab assignment. They are completely free and exemplify another indication of the support by JMP for the teaching and learning of statistics.

Figure 9.3 Modules and demos

INTERACTIVE TOOLS FOR LEARNING AND TEACHING

We want students to understand statistics. JMP®
interactive Concept Discovery Modules make it happen.

Nine modules that illustrate some of the most challenging and fundamental
concepts covered in introductory statistics courses, including the newest
addition, Demonstrate Regression.

The modules are fun, easy to use and available to professors, students and
any interested learner as a free download from the JMP web site. By
changing and controlling values, selecting radio buttons and otherwise
interacting with the module screen, users can see the concept in action
immediately.

Try out all nine to learn or teach fundamental concepts.

Note: These learning tools are part of the JMP File Exchange. Downloading of files from
the File Exchange requires creation of a sas.com profile.

1   **Distribution Generator.** Provides students with a more visual
understanding of the underpinnings of distributions and variability.

     ⇒ DOWNLOAD NOW   ⇒ WATCH THE DEMO

2   **Sampling Distribution of Sample Means.** Illustrates the distribution
of sample means and the connection to the distribution of sample data.
Includes controls for population shape, mean, and standard deviation,
sample size and number of samples.

     ⇒ DOWNLOAD NOW   ⇒ WATCH THE DEMO

3   **Sampling Distribution of Sample Proportions.** Illustrates the
distribution of sample proportions and the connection to the distribution
of sample data. Includes controls for population proportion, sample size
and number of samples.

You will need to sign up to get a JMP account, which is standard operating procedure for many websites. I place my experience with such sites on a scale from 1 to 10, with 1 meaning a neverending stream of commercial appeals and other messages, and 10 meaning that we're just keeping track of usage here, as part of our internal quality control. On that scale, JMP is a 9.9!

Almost all the modules come with a demonstration about how to use them. A teacher can run the demonstrations as part of formal instruction, and the student may use them as reminders when needed. My favorite is the confidence interval for the mean, and I will use that module to give a sense of what can be done with these modules. Of course, the dynamism will be lost in my description! If you follow along you will see this module in action. The confidence interval panel with my entries is shown in figure 9.4.

Figure 9.4 Confidence interval panel with values

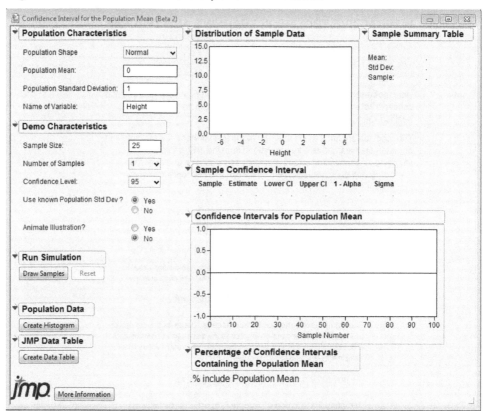

When I ran the simulation with 100 samples and selected **DrawSamples** I was presented with what is shown in figure 9.5. Of course, your results will differ from mine—such is life when simulating.

Figure 9.5 Simulation results

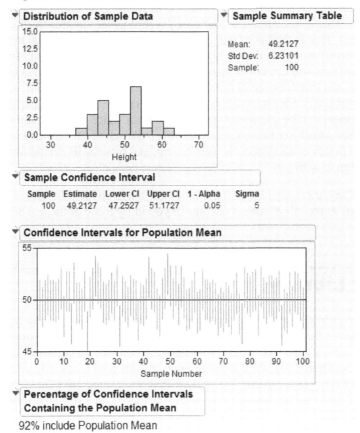

After running the simulation, your students may note the **% include Population Mean** is not equal to the confidence level, in this case 95 percent. You might point out that this is an example of what is known in the cognitive psychological literature as the Law of Small Numbers. That is the belief that if a probability of success is equal to 0.95, a run of 100 (or other small number) trials will give exactly 95 successes. Notice also that after the simulation stops you can either reset it or take additional samples. If students elect to reset, they will see that for a small finite number of samples, one does not always get the same number of confidence intervals containing the population mean. Variability at work! If they elect to take additional samples, they can watch the **% include Population Mean** get closer and closer to the confidence level, the statistical Law of Large Numbers in action. The power of simulation is well known, and these JMP simulations are not only informative, but easy to run.

As of this writing, there are nine of these modules, but keep looking—these will probably soon be known as the "first" nine. I encourage you to try more of these modules.

The Learning Library

Another resource at the JMP site is the Learning Library option shown in figure 9.6. This is a great resource for at least two kinds of people: (1) the instructor-scholar, burdened with many papers, students with questions, and other classroom paraphernalia; and (2) the assuredly less burdened but all-too-common student who reads only under duress. The Learning Library is a remarkably well-written set of materials. The remarkable part is not that it is well written, but that it is so well written with a space limitation of a single page per topic.

Figure 9.6 The Learning Library

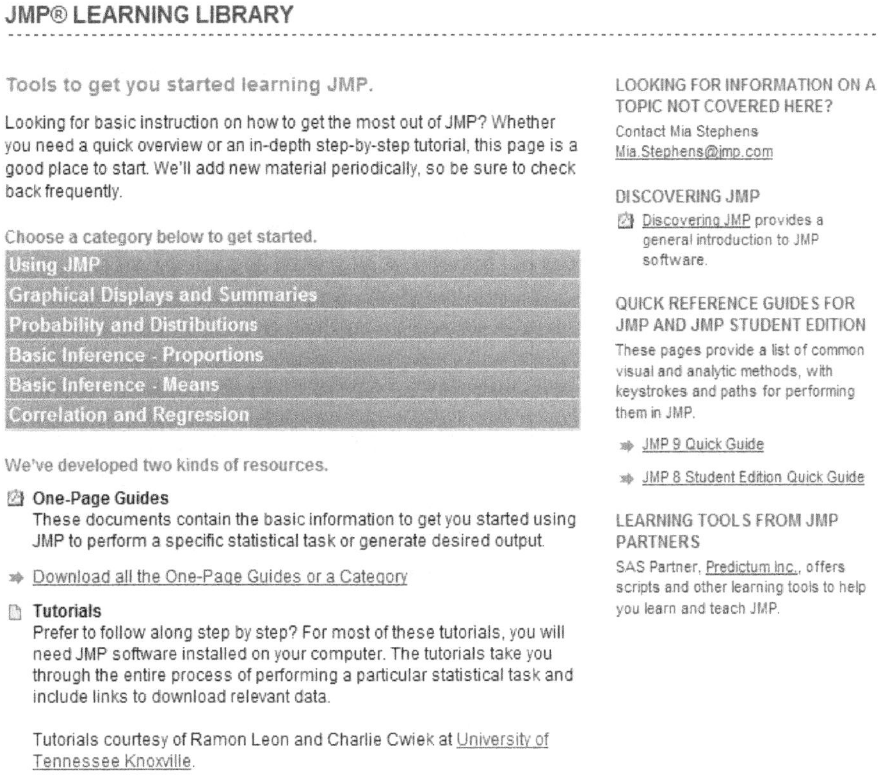

JMP® LEARNING LIBRARY

Tools to get you started learning JMP.

Looking for basic instruction on how to get the most out of JMP? Whether you need a quick overview or an in-depth step-by-step tutorial, this page is a good place to start. We'll add new material periodically, so be sure to check back frequently.

Choose a category below to get started.

Using JMP
Graphical Displays and Summaries
Probability and Distributions
Basic Inference - Proportions
Basic Inference - Means
Correlation and Regression

We've developed two kinds of resources.

One-Page Guides
These documents contain the basic information to get you started using JMP to perform a specific statistical task or generate desired output.

➡ Download all the One-Page Guides or a Category

Tutorials
Prefer to follow along step by step? For most of these tutorials, you will need JMP software installed on your computer. The tutorials take you through the entire process of performing a particular statistical task and include links to download relevant data.

Tutorials courtesy of Ramon Leon and Charlie Cwiek at University of Tennessee Knoxville.

LOOKING FOR INFORMATION ON A TOPIC NOT COVERED HERE?
Contact Mia Stephens
Mia.Stephens@jmp.com

DISCOVERING JMP
Discovering JMP provides a general introduction to JMP software.

QUICK REFERENCE GUIDES FOR JMP AND JMP STUDENT EDITION
These pages provide a list of common visual and analytic methods, with keystrokes and paths for performing them in JMP.

➡ JMP 9 Quick Guide

➡ JMP 8 Student Edition Quick Guide

LEARNING TOOLS FROM JMP PARTNERS
SAS Partner, Predictum Inc., offers scripts and other learning tools to help you learn and teach JMP.

The Learning Library, downloaded *in toto* as a .pdf file or referenced singly by topic online, offers a fast answer to the JMP user's simple questions. I must confess that the Learning Library is something I rarely use; the interface of JMP is so intuitive that even a small amount of experience gives one a kind of JMP "Zen." However, while I am making confessions, here is another one. I recently used one of these one-pagers. I wanted to analyze some data on Florida hurricane damage using simple ordinal logistic regression. Ordinal logistic regression is a technique I rarely have occasion to utilize these days, and as it happened, I was on the road without my JMP library. I brought up the "Simple Logistic Regression" page in the Learning Library to see what was there. Not only did I find the information I needed, but I also found something I knew JMP could do but I had never actually considered implementing with logistic regression: using different markers and colors in the display. To make a short story shorter, this one-page presentation was responsible for my graph being much better than I had originally envisioned.

To illustrate how to take advantage of the Learning Library, select **Graphical Displays and Summaries** to get to what is shown in figure 9.7, and then select **Bar Charts and Frequency Distributions** to get to that represented in figure 9.8. This is a topic we have already discussed; now I would like you to put yourself in the mind of a typical student learning to use JMP as well as statistics, and appreciate how much information has been packed into that single page. Even the most reluctantly reading student—perhaps *especially* the most reluctantly reading student—would find needed answers, compactly presented.

Figure 9.7 Graphical displays

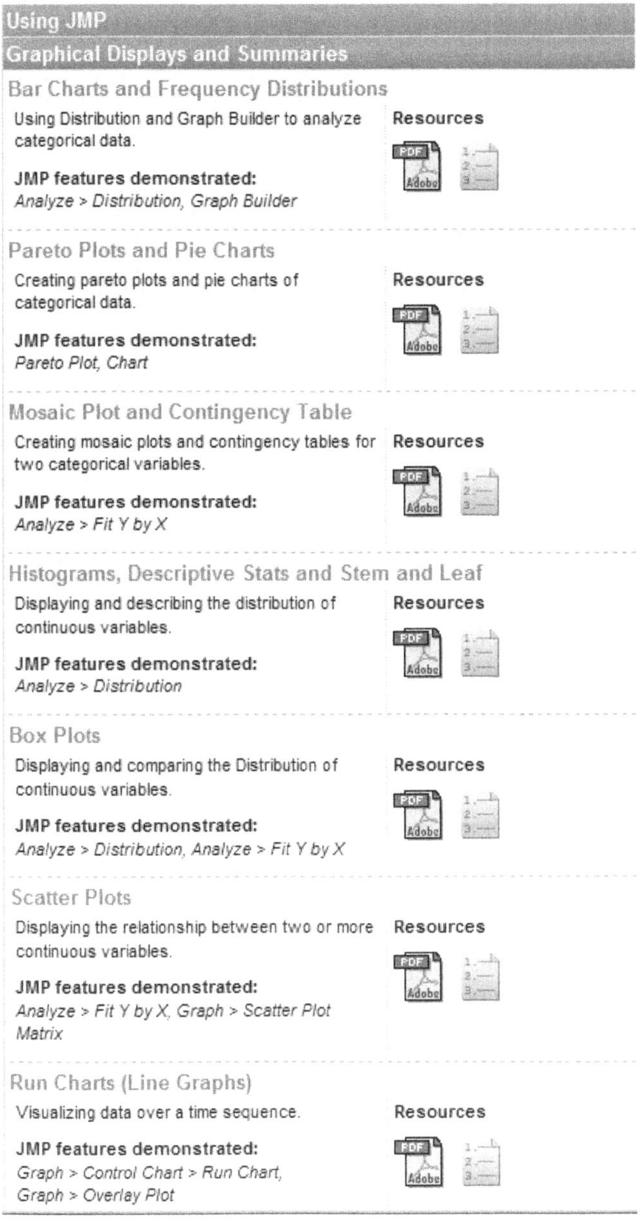

Using JMP

Graphical Displays and Summaries

Bar Charts and Frequency Distributions

Using Distribution and Graph Builder to analyze
categorical data.

JMP features demonstrated:
Analyze > Distribution, Graph Builder

Resources

Pareto Plots and Pie Charts

Creating pareto plots and pie charts of
categorical data.

JMP features demonstrated:
Pareto Plot, Chart

Resources

Mosaic Plot and Contingency Table

Creating mosaic plots and contingency tables for
two categorical variables.

JMP features demonstrated:
Analyze > Fit Y by X

Resources

Histograms, Descriptive Stats and Stem and Leaf

Displaying and describing the distribution of
continuous variables.

JMP features demonstrated:
Analyze > Distribution

Resources

Box Plots

Displaying and comparing the Distribution of
continuous variables.

JMP features demonstrated:
Analyze > Distribution, Analyze > Fit Y by X

Resources

Scatter Plots

Displaying the relationship between two or more
continuous variables.

JMP features demonstrated:
*Analyze > Fit Y by X, Graph > Scatter Plot
Matrix*

Resources

Run Charts (Line Graphs)

Visualizing data over a time sequence.

JMP features demonstrated:
*Graph > Control Chart > Run Chart,
Graph > Overlay Plot*

Resources

Figure 9.8 Bar charts and frequency distributions—on one page

Bar Charts and Frequency Distributions

Use to display the distribution of categorical (nominal or ordinal) variables. For the continuous (numeric) variables, see the page Histograms, Descriptive Stats and Stem and Leaf.

Bar Charts and Frequency Distributions

Example: Companies.jmp (Help > Sample Data)

1. From an open JMP data table, select **Analyze > Distribution**.
2. Click on one or more nominal or ordinal variables from **Select Columns**, and click **Y, Columns** (nominal variables have red bars, and ordinal variables have green bars).
3. If you have summarized data (a column with counts), enter the column into **Freq**.
4. Click **OK** to generate bar charts and frequency distributions for each variable.

Tips:

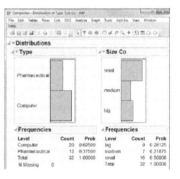

- To change the display from vertical to horizontal, click on the top red triangle and select **Stack**.
- To change future output to horizontal, go to **Preferences > Platforms > Distribution**, click **Stack** and **Horizontal**, then click OK.
- To change the graphical display for a variable, or to select additional options, click on the **red triangle** for that variable.

What Have We Learned?

In this chapter we considered two of the resources available at www.jmp.com: Interactive Learning Tools and the Learning Library. These are a subset of the resources of JMP that are particularly useful to teachers and students of statistics.

Index

X

Z

CPSIA information can be obtained at www.ICGtesting.com
Printed in the USA
BVOW10s0517230714

360166BV00003B/90/P